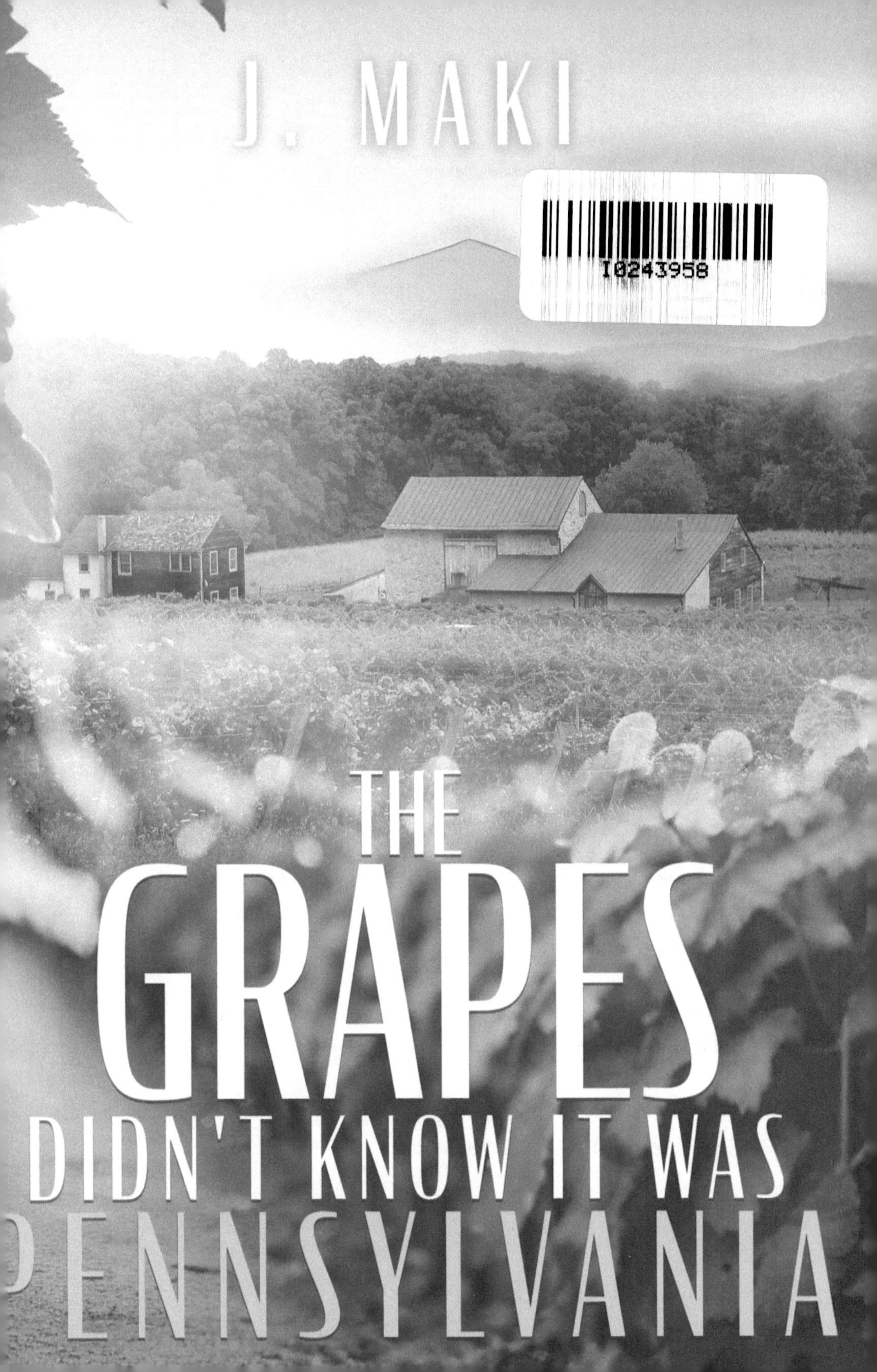

J. MAKI

THE
GRAPES
DIDN'T KNOW IT WAS
PENNSYLVANIA

SILVERSMITH
PRESS

Published by Silversmith Press–Houston, Texas
www.silversmithpress.com

Copyright © 2024 J. Maki

All rights reserved.

This book, or parts thereof, may not be reproduced in any form or by any means without written permission from the publisher, except for brief passages for purposes of reviews.

The views and opinions expressed herein belong to the author and do not necessarily represent those of the publisher.

ISBN 978-1-961093-40-9 (Softcover Book)
ISBN 978-1-961093-41-6 (eBook)

Many know the wine world to be a stylish place to hob nob with people in-the-know. The Wine Spectator magazine on the coffee table, vacations in Tuscany, Mediterranean Villas, dinner at the French Laundry in Napa, art shows and benefits, wine tastings on Martha's Vineyard, a drive in the country in an open-air Jaguar, sipping wine on a deck overlooking miles of groomed vineyards, family names that go back three generations.

This story is not about any of that.

In 2001, in Paris, France, at the prestigious Vinalies Internationales Wine Competition, my 1997 Blanc de Blancs was awarded a gold medal. It was the only gold that year for Blanc de Blancs—so for that moment in time, I had made the best champagne in the world.

People frequently asked, "How did you do that?"

The answer is that in a perfect wine grape growing season, I grew perfect wine grapes on a perfect wine grape site and made perfect wine. Something I never dreamed possible—after all—I was in Pennsylvania.

Perfection is only a moment in time. There are hundreds of factors, some small and some large, that go in to making the best wine; even the smallest misstep along the way can ruin the best wine grapes.

This is the story of that journey.

Dedication

This book is dedicated to the countless individuals
Who toiled in the vineyards over the centuries,
Who shared their knowledge,
and who mastered the art of winegrowing.

The Early Years

When I was seven, my family moved to Abington, Pennsylvania. We were five girls and one boy, but the boy was too young to pal around with. I was the one who liked to do things, make things, and find out how things worked. I liked all sports and wanted to play the boys' games.

I used our old Schwinn bike to get around to friends' houses and the school yard for volleyball games. I was ten that summer when I discovered the shopping center about two miles from our development, and the Sears store. I entered the Sears store into the hardware department and was immediately fascinated. I loved walking up and down the aisles looking at tools and equipment. I went there about once a week that summer. I was intrigued watching the salesman explain equipment to customers. In late August, the department manager asked me if I noticed that they were having a drawing for a prize. He showed me the table, and the prize was a new AM/FM radio. I told him I couldn't buy anything, and he said that was ok, and gave me an entry form to fill out. About six weeks later, Sears called and told my mom I had won the drawing. She took me to the store and questioned the manager as she knew nothing about this. He smiled and gave me the radio. I began to listen to various radio stations late into the evening, mostly from New York, and my world started opening up.

The next winter Mom took Nancy and me up to the Sharples Pond to go ice skating. Nancy, next older sister, wanted to visit with friends and I tried the skating. The pond had tree branches sticking through the ice in places. The boys were playing crack-the-whip. It was thrilling to watch. A couple of weeks later I set out from home with my skates to go back to the pond. I remembered the way; it was about a mile and a half. I started skating the edges watching for the boys. A few were playing ice hockey and after a while a group formed for crack-the-whip. I had asked earlier if I could join. "It's too dangerous," I was told, but I stayed for an hour watching until the shadows were setting over the pond. As the group started to break up, the boy I had asked earlier about learning to play came over to me and said that I could be on the end of the line next to him since half the boys had left. "Hang on tight," he instructed, as the line started to move. I was thrilled beyond belief as they swung me around again and again.

I took the American Red Cross swimming lessons at age eight and was invited to swim in a new team at Abington High School—the Abington Dolphins and was competing soon after. One Saturday there was a regional all day meet, and my coach asked me if my parents were coming to watch. I was surprised by the question as they never came to any activity I did. "No," I responded, "they were busy." Another time my coach sent a note home that Donna de Varona, the Olympic medalist was coming to an extra practice to see me swim. My parents never came to the practice.

When I was 11, I qualified to try out for the new Philadelphia Aquatic Club Team and to train for the Olympic trials. I made the team with Ellie Daniels, another girl from the Abington Dolphins. My parents said they couldn't get me to the practices, so my coach found a carpool, but they said no. This was and remains the biggest disappointment in my life. Ellie went on to the Olympics and I lived with knowing I always beat her when we raced.

We moved to Upper St Clair in Pittsburgh when I was 12 and I went to a public school. It was a shocking experience. There was very loose order, seemingly constant activity, and students moving everywhere. I was accustomed to the quiet discipline and politeness of the Catholic School. I wore my sister Nancy's hand-me-down clothes since my school uniform wouldn't work in public school. It was obvious to me that I looked different. A girl on the bus invited me to her house after school. Her mother was sewing a dress for her daughter, and I asked if I could watch. She answered my questions and explained what she was doing. I was amazed to watch her set a zipper in place. She served us a snack of milk and cookies and asked if I would like to learn to sew. I jumped at the chance and eagerly said "yes!" I was invited to come over the next day. When I arrived, her mother had picked out a pattern and suggested that we make matching shirts. It was a sailor inspired design of blue and white seersucker with a big collar with blue piping. I couldn't believe my good fortune and returned every day, paying attention to each detail. I wanted to learn everything. My friend's mother showed me how to set a zipper and make a buttonhole, letting me practice on scraps. My friend and I would coordinate to wear our shirts to school on the same day. I was proud of what I had made and thankful for this special friend and her mom.

The following year we moved back east to Yardley, Pennsylvania. The neighbor across the street saw us move in and asked if I was old enough to babysit. I said yes and would babysit every Friday and Saturday nights in the neighborhood. I saved the money I earned and when I had enough, I bought fabric and patterns and started sewing my clothes. I picked patterns that I could use at least twice, and the best patterns were for mix-and-match separates of skirt and vest. Soon I could make a skirt in one weekend. In 9th grade I made several skirts and vests. That summer I started to buy Vogue patterns as they were more stylish looking. For the start of 10th grade, I made a fashionable dress of bright colors, a jumper and a new skirt. I purchased two blouses and two pairs of trendy shoes. My first day at school was an eye-opener with lots of compliments. It was nothing like my first day in Nancy's hand-me-downs. I felt good about my efforts and proud of the results. A single mom that I babysat for learned about my sewing and gave me her very old Singer sewing machine, which I treasure and use today.

College was always expected. Dad moved out when I was 16, but our parents didn't tell anyone and pretended he was travelling more. Finally, in my senior year in high school, I was told there wasn't money for applications or college; instead, I was expected to stay home, take care of the house, and help with my younger siblings, since Mom was going back to school for her Bachelor's degree. She was an R.N. and met and married Dad just before the start of World War II. Now she wanted to pursue nursing again. My sister Nancy had just graduated from Duke University and I was making her bridesmaids dresses that summer. She intervened with Dad and Mom and suddenly I was enrolled at the Catholic University of America in Washington, D.C. In early December at my university, a young man on the third floor jumped out of the window committing suicide. We were told to arrange to go home immediately two weeks early for the Christmas holiday. There was a line for the pay phone and finally it was my turn.

Unbelievably, my home number had been changed and the new number was unlisted. What had happened? I tried again–and pleaded with the operator that this was my home–please call them; surely they would want you to! But no. I racked my brain and then called our next door neighbor Flavia Clarke. Flavia was always drunk and pushing her strange sons on my sister and me. One time I went to find Mom, and Flavia was showing off her sons dressed in girls' clothing, waving a liquor glass in my face. When our parents were not home, my sister Kathy and I would lock all of the doors and windows in our house and stay inside. Our instincts were correct. These two boys–Haddon and Brad Clarke–grew up to be the only two known serial killer brothers ever convicted.

Flavia did give me our new home number. When my mom answered the phone I asked, "Why did you change the phone number? And why didn't you tell me?" Her reply was a familiar "Oh well, we just did that." No explanation.

I didn't return to Catholic U. and helped with the housekeeping. I busied myself making curtains and reupholstering the furniture. I found a job selling Encyclopedia Britannica in Trenton, New Jersey, and was dropped off each day to go door to door in Trenton neighborhoods. It was a valuable learning experience—but very unglamorous.

I saw a notice in the Trenton Times for a music festival in Woodstock, New York, and the list of performers was extraordinary. An upperclassman at Catholic U was interested in music and had a car (key item at this time!) He finished classes in early June and could pick me up on the way. We purchased two 3-day tickets and planned our trip. Camping was expected. We gathered our gear—tent, cooler, cook stove, lantern, flashlights, water thermos and food for three days, maps and cash for gas, tolls and misc. We left Friday morning planning to arrive early and get our campsite set up before the opening performance. Somewhere on the NY Thruway we heard radio reports about a massive crowd descending on Woodstock. Finally, we were on the road outside of town going toward the venue. The roads were clogged. We saw signs on lawns offering campsites and decided to be assured of a site and walk the two miles to and fro the venue. This homeowner had four areas to camp. We set up the tent, left everything in the locked car and set out for Woodstock. We walked past two miles of cars and got into the festival area with time to spare. We had blankets to sit on, and found a spot just left of center stage. Saturday was an endless sea of people coming and going. The food and beverage vendors were overwhelmed. There were public announcements to locate missing children or individuals, directions for medical assistance, new locations for water and public toilets. They told us we were at least 250,000 and more were coming. Everyone was calm, patiently waiting, sharing with their neighbors. That night it started raining and many attendees started to leave. On Sunday we packed our car and walked to the festival site again with our ponchos and blankets. The fields were getting muddy, and the crowd was thinning. We wanted to stay to the end to hear Jimi Hendrix. We were rewarded as he played the Star-Spangled Banner and the Army helicopters flew overhead and dropped rose petals on the remaining crowd.

Later that summer I visited my high school friend Joe at Drexel Institute of Technology in Philadelphia. He convinced me to try Drexel, and after an interview they accepted me to start as a freshman mechanical engineering student. Dad took me to Trevose Savings and Loan for a school loan. Nancy had just finished at Duke, and he said he couldn't afford it with him and Mom separating. That was the official notice!

At Drexel I worked in the kitchen for two meals a day and had a hot pot for tea in my room with PB&J sandwiches. My summer jobs would pay for books, fees, and food. That first winter was the beginning of the college sit-in craze and the Vietnam War protests. I attended the March on Washington with a bus load of other students. I worked in the kitchen tent for the weekend and on the last day walked over the bridge to Arlington National Cemetery–a special and very somber place.

After the first year of calculus, physics and chemistry taught by the teacher's assistant I began to look for more. Pre-med opened up more science courses, especially biology and biochemistry, physiology, zoology, and microbiology. I took geology as an elective and was surprised at how interested I was in the earth's rock formations. I read Watson and Crick's famous book "The Double Helix" about the discovery of the structure of DNA. I decided to try to transfer to MIT. My SAT scores were very good–especially in math. They offered me a research assistant position to help with tuition. I was interested in genetics, but the costs were too great, and I couldn't get more student loans. So, I stayed in pre-med and took the Medical College Admission Test. The University of Pennsylvania Medical School was next door and would accept my application if I had a second signature for the loan. Dead end again. I had no husband and neither of my parents would co-sign.

After college, I was working in West Philadelphia at the Scheie Eye Institute as a research technician in glaucoma research and diagnostic testing and living in Bryn Mawr on the Philadelphia Main Line in a rented house. Dr. Vukovich was from Czechoslovakia. He taught me how to set up a laser for light studies and how to be very careful about it. Part of my job was to assist him in the patient diagnostic testing for glaucoma, as he had invented a technique to create a water bath around the eye and use an ultrasound to look for abnormalities. He drove a big Mercedes with bullet proof glass. One day he told me about how he came to America because he feared the KGB, and that they might find him here. A couple of weeks later he never showed up for work again.

Now that I had a back yard to use, I decided on a tomato garden. Since I am a Sears & Roebuck-for-life customer after winning the radio, I bought my first shovel there. So far, I had purchased a Craftsman metal toolbox, a spark plug wrench, a set of screw drivers, and an adjustable wrench. This shovel was my first gardening tool. And Sears also had tomato plants, Rutgers and Beefsteak. When I stayed with Grandma in Wooster, Ohio, that summer of 1963, my cousins would take me to the county fairs and their farms. Uncle Carl, Grandma's oldest son, was still farming on the family farm. I couldn't get enough of doing and seeing everything. I remembered their wonderful vegetable gardens and aspired to have my first.

I carefully surveyed the back yard and picked out a spot large enough for the twelve plants. There was a big tree in the middle of the yard, so I worked the area closer to the next-door neighbor. I worked the shovel into the dirt and started to dig up the grass. It was slow going and my hands were sore.

"Ahoy there," came a voice over the fence. I looked up to see my seventy-year-old neighbor. "I have some other tools you can borrow that will help." That was the beginning of a very special friendship with Whitten and Hortense Richman. Whitten showed me many gardening techniques for vegetables, bushes, and trees. Hortense loved flowering plants and shared her composting secrets. Whitten previously had a heart attack and had retired early from Abbots Dairy where he was an ice cream maker/chemist. He had taken up cooking and shared his recipes with me. We had a familiar joking about his cooking—he would ask me to taste his latest cookies to see what I thought. We would have long talks critiquing the cookies and also other things we had tasted before. We would argue about which tomatoes were best and the best tomato staking system. He showed me how to make a sour dough bread starter from scratch as good as any San Francisco starter! He gave me a wooden recipe box with some favorite recipes already written out. Each visit, I would ask for another recipe and we would discuss the ingredients and possible substitutions. He taught me how to substitute honey for sugar in a recipe. This turned out to be a lifesaver for me, since I was pre-diabetic, but didn't know it. I just didn't like sugar, and never consumed soda growing up. I began cooking with honey exclusively and sensed that it was better for me. Much later, I discovered that the honey molecule was different than the sucrose molecule.

Hortense was an avid reader and one day gave me her copy of *Ten Acres Enough: How a Very Small Farm Can Keep a Very Large Family* by Edmond Morris 1864. I read each page twice. This book resonated deeply. From then on, I dreamed of having my own small farm—ten acres and a barn. I never lost that dream.

My two friends from college, Joe and Tom, lived in a large house in Powelton Village with JT Cummons. And so did John Herbert, the son of Larry Herbert, Head of the Exhibits and Maintenance Department at the Academy of Natural Sciences in Philadelphia. Tom and JT worked for Larry. Another friend who worked for Larry was Italian and lived in South Philly. They had the family wine press that was brought over from Italy in their backyard, and no one wanted it anymore. John was interested in wine and got the press set up in their backyard on Baring Street. At the time, the Italians in Philly, NYC, and Boston would arrange for train carloads of wine grapes to be shipped from California. John arranged for Zinfandel grapes that year, 1973. I palled around with Tom and was just getting to know JT. Tom, JT, and I were all biology majors. Later on, I would tell people that in winemaking there is a lot of cleaning, so the guys were happy to have me work on this first crush. I still have a bottle of that first vintage and, as I learned later, Zinfandel takes a long time to smooth out! This was the beginning of a 50-year adventure. I loved making things in general, and it turned out that I also had a very good palette, which is a must in winemaking.

One day on the drive home from work, I thought I heard a 'knocking' sound from the engine. I was a devout Volkswagen Idiot Manual practitioner. Since my early Sears store experience, I would frequently stop by and purchase a few Craftsman tools. I started with the basics for an engine tune-up, and also a Craftsman toolbox. I was dating JT at this time, and we explored the engine sounds. We decided to pull the engine and set up shop in the garage of my rental house. Step by step we took everything apart, following the Idiot Manual. I spread all of the parts out in order, and carefully marked them. We replaced some older parts with new ones. I took the cylinders to a local mechanic Whitten knew to 're-groove' them. The rings looked fine, but I ordered new ones. We came finally to the crankcase–the heart of the engine. What to do? I wondered what was inside–and decided to 'break' the crankcase open. The crankshaft looked okay but we had it sent away for stress testing and some machining. We put everything back together carefully and held our breath. The engine turned over and purred!

Larry Herbert was a connoisseur of wines and introduced JT and me to French Champagnes in the early '70s. Sometimes after work Larry would send JT or Tom to the local state beverage store for a few bottles of Moet & Chandon or G.H. Mumm champagnes. It was early in my wine tasting career, but I knew immediately that these were extraordinary. That impression of soft delicate elegance on the palette remains today.

I planted a big garden and tried to grow everything. We were both interested in honeybees and acquired our first hive.

JT and I married in '73 and bought an abandoned three story 1890s Victorian brick twin on Summer Street in Powelton Village, Philadelphia for $1000. We were the first Caucasians on the block. There was one bathroom of sorts and the cold water worked. There was no heat and a full eight cubic yards of coal dust in the basement. Several old buildings in the area were open for salvaging. JT and Tom found an old gas stove and caste iron sink and installed them in the kitchen. We had to find wood for the stove for heating that first winter and lived in the kitchen. The backyard was full of trash and brush. I cleared it out, installed our one honeybee hive, and began prepping a spot for a garden. The honeybees were just fine in the little backyard, but we worried about the neighbors. After a few months, the neighbors started talking to us and the woman next door said she liked watching the bees.

This was the year we took my grandmother to the Hershey Gardens. Before she returned to Ohio, she came by surprise to see our abandoned house. It was a Saturday, and we were salvaging in a warehouse nearby. I had found an old drafting table, and we were carrying it home.

Grandma was standing on the step polishing the old door knocker when we arrived. I couldn't believe my eyes.

"Grandma, what are you doing here?" She replied, "I'm cleaning the windows in your front door". We ushered this 85-year-old lady inside. She wanted to see everything I was doing. She looked over every inch of the house and remarked how interesting it was and gave suggestions. She had come to encourage me! In the years ahead, I would think of her as one of my angels, always there.

The Philadelphia police department learned of our honeybees and asked if we could help with the spring swarm season. Honeybees swarm in the spring to make room in the hive. The queen takes off with half the bees in the hive and leaves a royal egg behind to become the next queen. Several times a week we received a call and would take our gear and our VW van to capture the swarm. One swarm was on the side of a car parked at the 12th and Vine center city police station, and many were in trees all over the city. That first year we captured 11 swarms.

One day, the swarm was huge and high up in a tree. We were called in the morning. After observing the swarm, we had to go for a bigger ladder and extra equipment. When we returned, we climbed the tree and positioned a box on the ladder top, and carefully brushed the main ball of the swarm to make it fall downward into the box. Next, we carefully placed the box on the ground under the tree. Many bees were flying, and they would come back to the tree and look for the hive. The queen's smell would guide them. We left, planning to come back before dusk. On our return, we found the open box much the same, but a small ball of bees was nearby. As dusk was coming soon, the bees didn't want to fly. We pushed the grapefruit size ball of bees toward the box opening. Then, suddenly there was some movement of bees from inside the box; many bees were walking to the edge of the opening. Honeybees have an advanced communication system. We watched in amazement as they organized themselves into a bridge. Using their little bodies they extended their legs front and back reaching for the next in line. They bridged nearly a four-inch space with their bodies, end to end, to reach the ball of bees on the ground.... and those bees walked "on the bridge" and into the hive!

Each year, we would make our way with Tom and his wife, Camille, to the Finger Lakes region of New York to pick wine grapes at harvest. There, we'd press the grapes and bring back the juice to make wine in our basements. In the Eastern U.S. table grapes, like Concord or Thompson Seedless, ripen in August. (Table grapes are the eating grapes available in the grocery store.) The wine grape harvest starts in September and continues till frost in the East. We would stay with John and Susan Herbert on Seneca Lake. John worked for Wagner Vineyards. John's friend Mark Wagner had his own vineyard and was experimenting with *Vitis Vinifera* grapes. This was mid-70s and it was commonly thought that these grapes of European descent, well known in Europe and California, couldn't grow in the East. These are the types of wine grapes (e.g., Cabernet Sauvignon, Merlot and Chardonnay) renowned around the world for making the best wine, but they are the hardest to grow. The truth is that no one had tried to grow them in the East since Thomas Jefferson and Benjamin Franklin brought *Vitis Vinifera* grape vines from Europe, only to have them die.

After Prohibition it was discovered that the root louse Phylloxera, which lives in Eastern U.S. soils, kills the European *Vitis Vinifera* grapevines. At this time American grape vines like the Concord, Catawba, and Niagara varietals were exported to Europe and inadvertently took the Phylloxera root louse with them. This root louse decimated many European vineyards.

That's when it was discovered that the American grapevines were resistant to Phylloxera. This started the grapevine grafting industry. The American grapevine was used as the *rootstock* (rootstock is what grows underground) and the European *Vitis Vinifera* grapevine was grafted on the top of the American root stock to create a European grapevine that would grow in the Eastern U.S. soils. Hence, the name *Grafted Grapevine*.

But Prohibition started, and it took until the 1960s for the vineyards in the East to start planting these types of wine grapes.

In the late 70s we began to experiment with Mark's early vintages, which were Cabernet Sauvignon and Chardonnay, known to us as "the King and Queen of wine grapes". John would give me tips and instructions each year on the basics, and I relished visiting his winery for an inside tour and tasting. The industry in the Finger Lakes was full of down to earth folks who loved to share information about grapes, winemaking, and equipment. We were always welcomed with open arms. They always celebrated their successes.

JT and I made our first champagne in the basement of our Powelton Village home. John had given us instructions. It worked! First bubbles!

We also loved to share our new wines. These small quantities were consumed before the next harvest. Now I would call them *'green wines'* (young wines without aging–like eating an apple that isn't fully ripened), as later I came to appreciate the power of time.

The grand old buildings nearby that had been salvaged were finally torn down. It was unsightly and unsettling to live near the vacant lots that were at the end of our street. One afternoon when I was walking my dog, I came through the half city-block-sized lot and a group of MOVE members with their pack of dogs approached. About ten yards away they separated. I thought they meant to let us by, but instead they encircled us. The dogs were all German Shepherds and were held on leash like fighting dogs. The MOVE (Philadelphia based Black Liberation group founded in 1972) members wore old unkempt uniforms. My dog stayed close by my side as I kept walking, staring at those in front of me. In the end, they parted, and we walked on, but I'll never forget the feeling of that attempt of intimidation.

A few of us on our two blocks, Winter Street and Summer Street, were interested in gardening and decided to start a community garden on these vacant lots. We just started and didn't ask! We scavenged boards and containers to set up spaces for individual gardens. There were five gardens to start, and the numbers grew each year. This was the first community garden in Philadelphia. We made a three-minute film about the garden and what it meant to the residents. It helped get support with the city administrators and we were left to use the lots for many years.

I continued looking for another job. My older sister Nancy had been hired by IBM from Duke University and had done very well. It was 1977 and there were still few jobs from the major corporations. Since the early 1970s there were very few jobs for graduates, and the economy was struggling with high inflation. The credit card rates were upwards of 18%. My first job after college paid $6800 per year, and JT made $4200. After paying the school loans, we had about $450 per month for food, utilities, clothing, gas, and repairs. Nothing was left for renovations, including heat.

I called the IBM office in Philadelphia but was told they weren't hiring. Sister Nancy offered to inquire and called to tell me there was going to be a little hiring nationwide. Over a six-month period, I called every week. Finally, the last time I called the response was "just a minute" and when they came back, they said "we are going to be looking for someone in the Financial Services Branch; could I come in for an interview?" In 1977, I started a wonderful career at IBM.

With this new job, we were able to put in hot water baseboard heat on the first floor of the house. I made curtains as room dividers to keep the heat where we wanted it. During the first years at IBM, we worked hard on rooms of this old house. We repaired, sanded, and refinished the floors; also, scraped and painted the walls. Slate was salvaged from the roofs of houses nearby before they were torn down. This was my first experience on a roof. I was scared but determined to control my fear.

We still used the salvaged gas stove with the wood box. One evening, my brother Paul and Mom came to visit. Paul had been hunting that day and brought a pair of cooked partridges wrapped in tin foil. We were going to a dance performance at the Community Center and put the partridges in the oven. My dog Mac (short for MacArthur) was an Irish setter, a bird dog. We came home to find the oven door open, and a neat pile of bones set on the tin foil, which looked to be carefully smoothed out on the floor. We looked around for a break in, but soon realized the only culprit was Mac, pretending to be asleep.

My early years at IBM as a System Engineer saw the inside of the banking industry in Philadelphia. My first assignment was at a branch operation's back office to do a time and motion study at Fidelity Bank. **I never forgot the lesson that it is more costly to fix a problem than avoid one.** I got to work on the first ATM installation in the U.S. at Girard Bank, and then on the first check processing machines to read handwriting. I was promoted to Marketing Representative and had a territory which included mid-size banks and insurance companies and law offices.

I had separated from JT the prior fall and was renting on City Line Avenue in Philadelphia. I met Fred Maki and we began to ride bikes together. I was beginning to train for my first Triathlon. I was a swimmer and bicyclist, but disliked running. Fred was a runner and a bicyclist. He was from Cape Cod and had surfed and fished all of his life. Over the next five years we learned scuba diving and took many trips to the Caribbean and Hawaii. We married in 1989 and Fred began his wine journey with me.

I was transferred to Northern New Jersey for my first management job in IBM. I continued to find my way to the Finger Lakes to pick grapes at harvest. In 1989, I attended a winemaking short course at University of California Davis. I took my 1988 Chardonnay with me for review by the two instructors. I learned many new things to improve my wines, such as fermentation nutrients, yeast strains, and the importance of understanding sulfur in the vineyard.

There were presentations on:

- Pinot Noir by Henry McHenry- Pinot Noir *clones* (yes, genetics plays a part in wine grapes. There are many different variations on each type of wine grape, and these are called "the clones")
- The importance of French Oak, tastings of barrel toast levels and different oak forests was discussed.
- Jed Steele from Kendall Jackson winery presented on Zinfandel and his favorite yeast.

- Pete Minor presented Viognier: A Rhone Variety.
- Mila Handley of Handley Cellars in the Anderson Valley spoke on Sparkling Wines.
- Merry Edwards of Merry Vintner's talked about the Viticultural region and their Chardonnay clone selections, and the difference between Barrel Fermented and barrel aged Chardonnay.
- Randall Johnson from Calafia Cellars talked about Cabernet Sauvignon and "cap management" during fermentation.
- Joe Collins of Briedand Vineyards talked about Riesling and cold fermentation in stainless steel tanks, and freezing the *must* (the pressed fresh juice of the wine grapes) to use in adjusting the sugars later.

I went home with visions of sugar plums in my head!

My medical research jobs had afforded me good lab techniques. I acquired simple home winemaking lab testing equipment for measuring sugar and acids, and more sophisticated tests for *malolactic fermentation*. In wine grapes, there are three main acids—malic, citric, and tartaric. *Malolactic fermentation* transforms the malic acid to lactic acid, which is much softer to the palette. This is an extra cost and somewhat tricky step in winemaking, but the results are worth it. The famous French white

Burgundies, referred to as Burgundian Chardonnay, are made this way. These tests could be performed without much investment.

In the late 80s I was promoted to a regional marketing management position in Northern New Jersey. My manager asked me to select a group of executives for him to host at an IBM Executive Conference being held in conjunction with the All-Star game that summer in Oakland, California. I was excited because baseball was a passion of mine. I started watching the Philadelphia Phillies with my dad when I was eight. I had my first taste of beer then as well since Dad would drink beer and eat crackers with peanut butter and watch the games on Saturdays. One of my swim friend's parents took me to a Phillies game at Connie Mack stadium when I was eleven. I remember walking up flights of stairs on the outside of the stadium to get to our seats, and the thrill of seeing a live game.

By this time, I already had the makings of a Mike Schmidt sports card collection. Recently, I had been to two shows looking for his rookie card. Northern New Jersey is a big fan base for sports cards, and these fans like the NY, Philly, and DC teams. I had almost 10,000 Mike Schmidt cards by now.

When the event was close at hand, my manager just happened to find one extra ticket to the festivities. I couldn't believe my luck. One of IBM's customer activities was an Old Timers Hall of Famers reception. I went to a local sporting goods store and bought a baseball in case I could get an autograph. The event was three days and the first night was the Old Timers reception in the hotel. I walked off the escalator into the mezzanine area and it seemed that I was early. I looked around and saw a small group of men standing toward the side of the room. There wasn't anyone else in the room. I cautiously approached them and realized they were the Old Timers Hall of Famers! They were chatting and smiling, and as I approached, they quickly turned toward me. Here before me was Warren Spahn, Harmon Killibrew, Orlando Cepeda, Bobby Doerr, and Ray Fosse. They asked what I liked about baseball. I talked about the Phillies, Mike Schmidt, and Steve Carlton and other current players. I said that I especially like to watch the fast infield play. I listened to them joking and telling a few stories. I felt awkward about taking up their time, and sorry that I didn't know more about each of them. Bobbie Doerr asked if I brought a ball to sign. It was nearly falling out of my bag. He took the ball and said, "How about it guys? Let's all sign!" They kidded as they passed the ball around. I graciously thanked them and left beaming with a one-of-a-kind memento; the signatures of four

Hall of Famers on one ball! At home I looked up each player. I wished I could have asked Warren Spahn about his famous quote: "Hitting is timing, pitching is upsetting timing". Still today, Warren Spahn holds the major league record for most all time wins by a left- handed pitcher.

For the 1989 harvest, I couldn't get to the Finger Lakes and decided to buy California grapes from a broker in New York. They had Cabernet Sauvignon and Merlot and I tried to use some techniques I had learned about at the seminar in CA earlier that year. As a home winemaker, things were limited. *Cap management* would be done in a trash can in the garage. This is when you work the 'cap' which is the skins that float to the surface and *punch down* into the liquid during fermentation. The skins hold a lot of color, and this motion helps the color to express out into the fluid. I found a small fruit press with a hand operated wood crusher at a local flea market. This was my first solo pressing.

The list of supplies was short:

- Crusher to 'crush' or gently break the grape berries from the stem
- Clean trash can for the crushed grapes. Yeast to ferment. Paddle to punch down the cap.

- Big funnel to catch the liquid poured from a bucket into the glass carboy (5-gallon glass jug).
- Pail to transport the grapes from the trash can to the press (when finished fermenting), and again from the press to a glass carboy.
- Potassium metabisulfite to prevent unwanted microorganisms.
- Wood chips to add to the wine for aging.

We managed to get the fermented juice pressed and fill two 5-gallon glass jars called carboys. The carboys were kept in the basement at about 65 degrees F. That winter, I racked the juice off the sediment and bottled two cases each of the Merlot and Cabernet Sauvignon using a hand-held corker.

For my "day job" I had been teaching in a business school IBM started with Harvard and MIT on evaluating investments in Information Technology for U.S. companies. In 1990, IBM started the IBM Consulting Group, and our organization became the founding members. We were being assigned geographic territories and I was asked to move to Los Angeles. Since moving from the Philadelphia area to the New York area, Fred had tried several jobs and was currently working as a sales rep for a NY regional security systems company. If we could locate a place to start a small vineyard and winery, Fred would become the farmer, and was looking forward to it.

By now, I had seen enough of the Finger Lakes wine region in NY to know that the small winery operations were appealing; I wanted to start on this endeavor. At the time, California vineyard land was at least $20,000 per acre. The New York region would be a natural fit, but the territory was taken. Washington and Philadelphia were open, so I began a search for a site within an hour of the airport, since I flew often for work. I researched the characteristics of great vineyards around the world. The University of Virginia identified a swath of land from Virginia, across Maryland, Southeastern Pennsylvania, Northern New Jersey and Long Island which was referred to by the author as a Gold Coast for European wine grapes.

After a few months of analysis, the best bet seemed to be southeast Pennsylvania. We started an intensive six months scouring the counties in southeastern Pennsylvania and finally centered on Chester County which was known at the time for its commitment to preserving farmland. Every weekend for six months, we drove the roads of Chester County. Four times we drove on Grove Road past an old farmhouse and barn that looked unused. The owners were deceased, and it took a bit of time and luck to find the relatives. They were about to sign an agreement with a builder but agreed to meet with us the Saturday before they signed. The small stone house had a 1700s log cabin as the kitchen. It had a sink, and the floor was dirt with a piece of linoleum under a table. No other appliances were in the room, and none of the plumbing worked. It reminded me of the abandoned three-story brick twin in Philadelphia that JT and I had bought twenty years earlier. The barn was 1820s with an inch of concrete holding the stone together. It was a dairy barn with remnants of the stalls, and the upstairs hay and tractor loft floor was still in good condition. The land was a south facing slope, perfect for a vineyard, and we knew the soil type was Chester County schist and very rocky, very good for wine grapes. We sat on the shed porch step outside with the owners. They loved our idea for the place, and we agreed to a sales price, contingent on the soil sample results. We signed on a paper napkin.

Route 23 starts in Philadelphia and ends up in Lancaster, Pennsylvania. The road follows a ridge line. In Chester County the land north of Route 23 is more clay, and south of Route 23 is granite and schist. Today if you drive west, you will see the stone color on the north side is reddish from the clay, and on the south side it's tan and gray with mineral streaks. Grove Road is south of Route 23 and the top of the vineyard is the highest point for ten miles. The soil is what is called "well drained". Water runs down quickly away from the surface, and the higher elevation means the cooler air flows down also. Wine grapes don't like their "feet" wet and the good air drainage helps in the late spring when frost might be a factor. We knew something about all these things, but only time would tell how significant this came to be.

That was the winter of 1991. With four months till the property closing, I called the grape grafters in the Finger Lakes to see what might still be available. Herman Amberg and Hermann Wiemar were trained in Geisenheim, Germany, and were thought to be the most expert grape vine grafters in the U.S. at the time. I knew them both from my many visits to the Finger Lakes. Hermann Weimar was also a winemaker and made incredible Riesling.

Usually, a grafted vine was ordered a year in advance. The grafting took a year to take hold. There were many rootstocks to choose from. The SO4 and 3309 rootstocks were doing well in the east, not too much vigor for the vine, but enough. I scouted out about 40 Chardonnay plants and 350 Pinot Noir plants.

1991

We closed the purchase on April 5, 1991, the first Friday in April and immediately went to the new property, our Dodge Caravan loaded with camping gear, tools, buckets, and the boxes of vines. The thirteen acres at 200 Grove Road, Elverson, Pennsylvania would never be the same. The vines were immediately put into buckets of water to rest for the night. We carried our camping gear and clothes into the log cabin bedroom. We had a Coleman cook stove and lantern and sat on the front porch floor to make our cookout of hot dogs and baked beans. Our cooler was full of Corona and lime. There was a saying I learned through the years of grape harvest work in the Finger Lakes—"It takes a lot of beer to make a bottle of wine." We had peanut butter and jelly for sandwiches, a half-gallon of milk, coffee for breakfast, and made toast on the Coleman stove.

We planned to dig the vines in by hand. The roots were about 20 inches long. The instructions are to dig a hole deep enough for the bottom of the rootstock, and to spread the roots out separately like ribbons from a Maypole in the bottom of the hole. By noon on Saturday, we had five plants finished. We sat out of the sun inside the old milking shed on the dirt floor on some scavenged wood and ate our PB&J sandwiches. After lunch, we continued till sundown and had 15 plants completed. We drove out for more sandwiches and then climbed into the sleeping bags in the Log Cabin bedroom. We resisted talking of how tired we were, with so little done. In the middle of the night something woke me. I had this feeling something was in the room watching us. I lay still and tried to get my night eyes working. The startled feeling left me and was replaced by a friendly feeling. I told this to the property relatives the next day when they came to visit and they said, "It must have been Percy checking you out. He must have decided that you were ok." Percy and Emiline Yost had lived here for the last 40 years and died in that bedroom.

On Sunday, we were determined to finish the Chardonnay row of 40 plants. We started early digging each hole. I would put the plant in place, push the dirt around the plant, with a final tamping around the plant with my boot. By noon, we knew we couldn't finish, but had Monday morning before we travelled back to northern New Jersey. By Sunday night, we had 33 plants in, and we were sore all over. The next morning, we decided to secure all the remaining plants in water and out of the sun. They were inside the old dairy barn and would hopefully be safe till we got back the next weekend. We recovered during the week and tried to think of ways to improve our work but were unsuccessful without an investment in equipment.

The next weekend we finished the Chardonnay plantings. Fortunately, the Pinot Noir vines were smaller, and the roots shorter. This meant the holes didn't have to be so deep, and we made better progress. The following weekend it rained, and the soil was easier to dig. We were getting better at our task and had figured out a few things to move along faster. We kept at it each weekend through May. Meanwhile, we found a house to rent nearby for four months, while we employed a contractor to make the old farmhouse livable.

During these weekend trips from northern New Jersey to our new property in southeastern Pennsylvania, we would have our road bikes with us to continue riding in preparation for the Denali expedition we had planned over 18 months before. During the last five years, we had completed a mountain climbing course on Mount Rainier with the Rainier Mountaineering Guides and also a Mexican Volcanoes climbing trip with them. We were each training weekly to be ready for this adventure. I was swimming a 2-mile interval workout with a Master's coach Monday through Friday and riding about 100 miles a week. The Rainier guides said that swimming is a perfect training exercise for mountain climbing, since breathing is core to the climbing effort. They had pioneered a breathing teaching approach they called 'pressure breathing.' Basically, it's simple—as you take one step up hill you breathe in, and on the next step uphill you breathe out. This forces a slow steady approach that everyone can do.

Fred was working out in a gym, and we had been riding bikes together in northern New Jersey. Here at the farm in Pennsylvania we were on a steep hill, and Fred had some trouble with the climb. I didn't think much about it, and he said he was just out of practice. Each weekend, he would decline to ride the hill. Finally, he said he wouldn't be ready for the Denali trip. The trip required climbing partners—a safety measure—so I couldn't go alone. After more prodding on my part, I realized that Fred wasn't comfortable for some reason, and we cancelled the trip. We stored all of our gear till next year.

In 1991, I started keeping a vineyard journal and recorded the purchase of our first vines, grape variety, clone, rootstock, nursery and grafter, and placement in the vineyard. Each year I would add to this vineyard map, keeping the grafter's tag of rootstock and clones. As the years went by, I could determine differences in growth and wine flavors.

June first was moving day from northern New Jersey. The moving company left the Allendale house the day before and placed everything in storage for four months.

Weekend days were spent at the farm working on the vines or the house. During the week, I was driving to Wilmington, Delaware, each day for a project at First USA Credit Card Company. I would leave at 6AM and stop at the Brandywine YMCA to swim, and then proceed to Wilmington. Our project was to build a new Customer Service System.

If there was daylight after work, I would go to the farm property on the way home and check on the construction work. Fred oversaw work on the property and kept track of the contractors on weekdays. On weekends, we would care for the young vines and work on scraping, repairing, and painting the old house plaster walls.

I was a member of the American Wine Society and located a regional chapter. Over the years before starting our farm, I had completed their Wine Judge training and participated in many wine tasting events. This year

I decided to enter our 1989 Cabernet Sauvignon in the annual competition. It won gold and best of show.

Common thinking is that it's the winemaking that makes great wine, but it's actually the grapes that make great wine. You must have the best grapes to make great wine. Continuing that thinking, we realized that we must focus on growing the grapes to make sure they are the best; especially, if you are a small winery. A single "off" tomato in a big batch of sauce won't make a difference, but one "off" tomato in a small batch will decline the result. Every cook knows this.

So, we went the slow road by planting a vineyard from scratch and were prepared to wait three to five years for a crop. With wine grapes, it takes five years for a full crop. We were 40 years old when we started the vineyard and thought nothing of waiting. My friend, Mark, said that this was the best way to start a farm business; because if you grow a crop, you can sell your crop on your property. I later learned why this was so important when I was applying for the winery licenses.

On September 30th we had vacated the rental house and were camping out at the Grove Road farm. The moving company was due on October 2nd. The house had a new kitchen with the 1840 pine floorboards from the attic, and re-use of the dairy stall boards from the barn. We had added a small addition of one room downstairs and one bedroom upstairs with a full bath. The first recorded deed of the property was in 1765 through the Port of Philadelphia.

It's important that the initial planting does everything to get the vines off to a good start. We planted all our vines in the ground by hand in eleven rows. At the end of that first year, we knew the property had good soil and the vines would grow well, and realized we would need equipment sooner than we thought.

1992

We bought a used John Deere 950 diesel tractor (circa 1970) with a used five-foot brush mower that winter. This would be the work horse for mowing the total property of 13 acres.

In 1992, we also started the winery application process. In Pennsylvania, wineries are controlled by the Pennsylvania Liquor Control Board (PLCB), an entity started after Prohibition and one of the two remaining state liquor boards in the U.S. The other is Utah. The PLCB is the single largest purchaser of alcoholic beverages in the world; in 2020 their annual sales were $2.56 billion. Today, there are over 600 PLCB or 'state' stores that sell alcoholic beverages including wine, which are direct competitors to the more than 120 Pennsylvania wineries. Therefore, financial success for a Pennsylvania winery is very difficult.

Pennsylvania was founded by William Penn, a Quaker. Alcohol is frowned upon in the Commonwealth, despite the State's business in it. But beer is thought about differently; currently, there are over 300 breweries in Pennsylvania. The general Quaker theme is that everybody is the same; hence, mediocrity thrives. Pennsylvania enjoys a nationwide reputation of thriftiness, where innovation is not the norm. I also knew from my early IBM days in Philadelphia that outsiders were frowned upon.

Our first step before filing for the winery applications was to file the name with the Pennsylvania Department of State. The French Creek streams run nearby, and we are on a ridge line. We decided on French Creek Ridge Vineyards. Having 'vineyard' in the winery name means you have your own vineyard.

Once that name was filed with the Pennsylvania Department of State, I proceeded on both the state and federal winery license applications. The state application required the approval of our Township. This is when it made a difference that we had started the vineyard. At the Warwick Township meeting, I presented the winery application to the Township Supervisors. They talked in front of me about how they didn't want a winery in the township. However, because we were already in agriculture with the vineyard, they couldn't refuse the application.

Also in 1992, I purchased 500 Vidal grape vines from Boordy Vineyard in Maryland. Their Vidal had won me over in a tasting and was thought to have come from more Chardonnay in the genetic crossing. Vidal was a *French American Hybrid* (genetic cross of American grape vines and European grape vines) and the only hybrid that we planted. I wanted to make Champagne and would use this to experiment.

French American Hybrid grapes were popular in the East because they were easier to grow than European grape vines, and more flavorful than the native American grapes. Hybrids were a cross between American grapes like Catawba, and a European grape like Chardonnay.

Hybrid grapes are "own-rooted" and the rootstock is the American rootstock. They don't have to be grafted. They are less expensive per plant and yield almost twice the crop per acre than the European grapes.

The hybrid vines are smaller and easier to plant ordinarily, but the area I picked for our Vidal was extremely rocky. We had planned to dig these in, but it was impossible. We hired a local farmer with a posthole digger to dig holes. I would come home each day from work that spring and plant 25 vines. The 500 vines took me about three weeks.

Meanwhile, our first-year plantings made it through the winter. They grew like crazy in the 1992 growing season

My project in Delaware ended that summer and I was now commuting to PNC Bank in Pittsburgh on a Sunday night or

Monday morning flight, and back Friday on a late afternoon flight. Fred kept up the mowing and hand work around the young vines and was painting the windows. We had yet to paint the inside of the house, but the heat and electric worked; but still no A/C and no downstairs bathroom.

My first dog passed away in the early 80s and I had been hoping for an opportunity to get a new puppy. I met two dog owners on a business trip back from Colorado that winter. They were talking about the dogs on board while we were all waiting to deplane. The dogs were a hunting breed from a breeder in Reading—Field Red Setters. After visiting the kennel, I fell in love with the puppies and that June picked out a female. We named her Tashie. She delighted us all summer following us to and fro. I set up an umbrella next to me working in the vine rows while tending the young vines. I couldn't wait to get home Friday nights to see this bundle of joy.

In August of 1992, we estimated the crop on the two-year vines—it was very small as expected but we wanted the chance to taste wine from our first grapes.

The harvest equipment checklist was the same as in New Jersey: the small fruit press, a crusher, and the various containers.

When the time came, we harvested Pinot Noir by hand and had about 5 gallons. I had only Pris de Mousse wine yeast on hand. This yeast is a work horse in wine making with very reliable results. This was the wine yeast used for our first Zinfandel made after college. After this vintage, I tried at least one new yeast per year. It would be at least two years until we had any idea about the quality of this first Pinot Noir wine.

1993

The winter was marked by a major storm in March. Power outages were common since we moved in. Our road was the 'end of the line' between two electric companies. We were the last to get service restored. The power outages were a break from civilization, and we enjoyed the quiet time.

This year would yield a small harvest of three-year-old Chardonnay and Pinot Noir, and the first vintage that we would bottle to sell.

After *bud break* in early May 1993, we had an idea about the quantity of this first crop.

The fruit on a grape vine grows on the shoots that grow from the canes that were produced in the last growing season. Hence, the phrase "you only get grapes from last year's wood." The *buds break* is in the spring, similar to other plants like flowering trees that are said to be *budding out*.

With an expected first vintage, I started a new winemaking journal since this would be my first commercial harvest.

This was my first entry:

May 18, 1993—First row Colmar clone Chardonnay, not enough crop to press. 350 Pinot Noir, use the fruit press, get one new oak barrel, and try Epernay wine yeast.

I started researching wine barrels and decided to try World Cooperage by Independent Stave in Lebanon, Missouri. They used American white oak from the Eastern US. Eastern white oak is considered "tight grained." My friends in the Finger Lakes recommended World Cooperage. I ordered one Bordeaux barrel, 225 liters (59 gallons), medium toast. There are many variables in wine barrels, which I would learn more about in the coming years:

- Shape of barrel - either Bordeaux or Burgundy style
- Age of the wood *staves* (strips of wood used to form a barrel)
- Oak forest—In France, there is an educated industry to maintain their famous oak forests for wine and cognac barrels: Limousin, Nevers, Troncais, Vosges and Allier. I would try them all over time.
- Toast level—light, medium, medium plus, and heavy. Barrels are fire charred on the inside. This brings out more subtle flavors from the wood.

I'll never forget the excitement when that first barrel was delivered by UPS. I used a wood center punch to mark the date into the front end of the barrel near the cooper's mark and added an entry in my wine journal. The cooperage instructions said to rehydrate the barrel prior to use. I filled the barrel with cool water and added a 10% sulfite solution. Each day, I would check the barrel and top it off with a little water. This process took about a week to complete. Just before filling, the barrel would be emptied and drained overnight. In later years, I would learn about burning Sulfur candles inside the barrel; this uses up the oxygen in the barrel and makes it sterile for a time.

I was becoming familiar with the local hardware store, Elverson Supply, and their friendly family atmosphere. They were interested in what we were doing and always happy to help with ideas or special requests. They cut 4 by 4 southern yellow pine sections to make a barrel stand. I used wood blocks to keep the barrel from rolling side to side; not fancy, but it worked.

Fermentation locks from my home winemaking days fit the barrel perfectly. This simple device allows the gas from fermentation to escape without letting anything in. During fermentation in wine, the wine yeasts "eat" the sugars in the wine juice and convert it to alcohol. Carbon dioxide (CO2) is the byproduct and is released through the fermentation lock. Unless the fermentation is *stopped* (a fermentation is *stopped* by chemical addition or cold shocking), all the wine juice sugars are used up in the fermentation. In general, wine that is sweet has sugar added back in to taste.

The human threshold to detect sweetness is 0.5% residual sugar (R.S.). This means that in any liquid the consumer will think there is **no** sugar in the beverage if the R.S. is 0.5% or less. I learned later that the average sweetness added to wine is 0.75% R.S. That amount is usually thought as the fruitiness from the grape. This is a confusing concept for the consumer. Later I offered our customers a tasting of R.S. levels so they could understand their personal preferences.

We first fill the barrel allowing for *head room*–space in the barrel to allow for the fermentation to occur without overflowing the barrel. After the fermentation is complete the barrel is filled to the top, called *topping off*, I carefully rehydrate the yeast according to the manufacturer's instructions. (Heat distilled water to no more than 105 degrees F, add the yeast, stir, cover from light, and wait ten minutes). Then, the warm yeast solution is added to the wine juice–in this case, into the barrel. Stirring is helpful, then the fermentation lock is placed in the bung hole of the barrel. Within a few days, the fermentation has started. The slow sound of the gas escaping through the fermentation lock always sounded like soft percussion music to me, and I always loved entering the room and hearing these special sounds.

During the late spring and summer that year, I began to clear the road banks of debris; broken glass, barbed wire fencing, road debris and other trash. I would take my lovely new garden cart with me out the driveway onto Grove Road to clear the lower bank. I had noticed a sign on RT 23 for *Daylilies* and decided to plant the bank with them. After I had cleared about 30 yards of bank, I was ready to plant. During the fall, I planted 30 daylilies on the lower bank in hopes of flowers next June.

At the end of December, we received our Pennsylvania winery license. Next, we had to register with the Pennsylvania Department of Revenue and continue working on the Federal Application, which is handled by the Bureau of Alcohol, Tobacco and Firearms.

1994

We continue working the first little vines planted in 1991 to encourage growth. It's time to put in the *trellis*, the support structure that holds the vines upright. The structure is made up of 5-inch diameter, eight-foot-tall southern pine pressure treated posts for the "line" posts, every 24 feet. The end posts are 7-inch diameter and 10 feet tall, set at an angle to cantilever the 22-gauge steel wires that will hold up the mature vines with their crop load. Our goal this year is to get the posts installed and the first wire - which is 30 inches above ground. This wire is the first support wire for the vines' trunk. We'll put the other wires up as the plants grow taller.

We also planted 600 Cabernet Sauvignon, 200 Cabernet Franc, and 500 Gewurztraminer plants. The two Cabs (short for Cabernet) were for making a Bordeaux Blend in the future.

Our wonderful old John Deere tractor now had a post hole digger attachment. I developed an approach to stake out the row with 2-foot wood stakes, string a line, then band under the line with weed spray. We had enough stakes and line for ten rows. When the weed spray result was evident, we would move to the next area to prepare for planting. Fred would work the tractor, and I would eyeball the post hole digger placement and depth of the hole. We would dig holes one row at a time, and then plant the vines.

A local farmer had a work shed full of thirty-inch rebar and offered it to us. We decided to try it for staking the vines. The wood supports with the first planted vines didn't fare well and had to be replaced. We were advised against using rebar, as it might leach iron into the soil, however we decided to give it a try.

This fall we would have our second harvest from the first plantings and the first harvest from the Vidal. I was determined to try a champagne. Everyone said. "Don't try that first!" I would ask "Why?" and the answer would always be, "because it takes so long to learn how to do that." Then I would answer, "That's why I'm going to start now!"

Champagne is a place in France and also the name given to a special wine made by way of a multi-step process called Methode Champenoise. Champagne is an English name and was the name of the lord who owned what is now the Champagne region in France.

Winemakers who make this special wine refer to it as *Champagne* if it is made in the *Traditional Methode*, which includes only grapes of Chardonnay, Pinot Noir and Pinot Meunier. This is confusing by design. There is no patent on the word Champagne, but many would have you think that. The French want to lay claim exclusively and are very good at marketing. There is another way to get bubbles into wine: pump CO_2 into the wine like soda is made. The word 'sparkling' is used in many countries (including the U.S.) and describes all carbonated wines–including Methode Champenoise, and CO_2 added sparkling, more confusion.

My aspiration was to make Blanc de Blancs, (French for "white of the whites" made from Chardonnay) and Blanc de Noirs ("white of the blacks" from Pinot Noir) in the traditional Methode Champenoise process.

I loved my visits to France and the meetings with French winemakers, wine growers, equipment manufacturers, and the people. I believed they were a century ahead of the world in the development of the wine grape industry. I aspired to make French styled wine. A Burgundian style Chardonnay was first, and then to try to make Champagne. This year would be my first attempt at both.

I had purchased my next two wine barrels–one for the Burgundian Chardonnay and one for the Pinot

Noir—both of American oak. Both barrels were from World Cooperage, but with different toast levels. This year I would try Epernay yeast, and a yeast nutrient called Superfood from the Wine Lab in California. One reason for *stuck fermentations* (when the fermentation doesn't complete) is because of lack of nutrients in the wine juice. Adding nutrients is like taking vitamins; very expensive, but less expensive than trying to fix a stuck fermentation.

Superfood is a mixture of yeast hulls, diammonium phosphate (DAP), yeast extract and vitamins. You can buy these items individually, but over the years this product will prove its weight in gold. It will become very important in the secondary fermentation process of making champagne and I believe it contributed to our champagne style of creamy, yeasty aroma. In addition, over almost 30 years we never had a stuck fermentation.

This year, we would purchase our first commercial wine press—a ½ ton bladder Idapress from Italy through Presque Isle Wine Cellars in Northeast Pennsylvania. Either compressed air or water could be used to expand the bladder. We choose compressed air and were off to Sears for an air compressor and hoses. We planned to use the little fruit press crusher again and bucket the crushed grapes into the press.

This year's harvest list was:

- Idapress
- Two barrels from World Cooperage, light toast and medium toast
- Epernay yeast and Superfood.
- Air compressor and hoses
- New outlets in barn
- Barrel stirring wood stick—used in the process of making Burgundian Chardonnay

Meanwhile, the Feds were coming for a premises inspection. They stayed for a very long time asking for more personal information. We were relieved when they left. In about a month, we got our Federal Winery License and were now officially in business.

We set up the new wine press in the dirt paddock area on the front side of the barn. The air compressor was from Sears, along with air pressure hoses, wheelbarrow, and various buckets. Just before harvest we bought some used equipment: an old dairy tank, Waukesha stainless steel dairy pump, and 55-gallon food-grade plastic containers from a Lancaster County farmer. I ordered new hoses for the pump. We bought used picking lugs (40-lb plastic fruit containers called yellow harvest lugs) and were ready to go—our first harvest!

The half-ton basket press had two large and heavy wooden baskets, held together with metal rods, on a solid steel base. Using compressed air, the press had to withstand at least three atmospheres pressure. Fred positioned the two basket halves securely with the solid metal rods. After a half hour he asked me to check the air pressure to make sure it was holding at three bars. When I bent down to look at the pressure gauge and adjust the air, the solid metal rod just above the pressure gauge flew by my head, missing me by an inch. It could have caused a major injury.

After cleaning, we set the press basket halves together. I watched Fred put the metal bars in place. Each one has a locking mechanism. I started double checking everything.

We picked the Chardonnay in the morning and pressed in the afternoon. It went slowly. There was a lot to clean up with emptying the press, double washing everything and taking the press remains to the compost pile using our Sears wheelbarrow—everything by hand at this point.

We pumped the liquid inside the old dairy barn. The floor was dirt and generally old and dusty. I hung sheets of plastic on the ceiling and around a 20-foot square area, and covered the dirt floor with a tarp, to make as best as possible a clean area. We had two containers that year with wine. The temperature dropped quickly before the *fermentation* (fermentation is the process where the yeast eats the sugar in the wine juice and converts it to alcohol) was complete. Wine fermentations at harvest time usually take 10 days to 2 weeks. The yeast is best if it is rehydrated in warm water; about 10 minutes at no more than 105 degrees F, then added to the wine grape juice—called 'the must'. This is the best way to ensure that the fermentation starts quickly. Fresh grape juice will begin to develop *off* flavors (both smell and taste) quickly, so getting the fermentation started as soon as possible is very important. **Half of what you taste is what you smell.**

The yeast will slow down below 50 degrees F. I wrapped heating blankets around the containers and put tarps over them. One day when I checked the fermentation status, I notice a small bat was clinging to the heating blanket. I decided not to disrupt him. The bat would earn his keep in the growing season eating bugs all night.

After we received our Federal License, we moved the wine into the barrels in the new space, ordered a small tabletop bottle filling device, and picked up several dozen sterile wine bottles from Gamber Glass in Lancaster. They also sold honey supplies and I had been there for honey jars last year for my one hive of Italian bees. We ordered corks from Presque Isle and would use my hand corker for now.

We would bottle the very small first vintage from 1992 that winter and open for customers. I hand painted a board with the words "Winery Open" and an arrow pointing down Grove Road and tied it to the guardrail on RT 23. On Saturday and Sunday afternoon, we would listen for the crunch of tires on our stone driveway and delight in getting a few dollars for a bottle of wine. Now we were Open for Business!

1995

The big project in the vineyard this year is to plant 2500 Chardonnay vines for future champagne. I selected the Colmar *clone*. Yes, a clone. Genes are involved in wine grapes. A clone for a grape vine is selected plant material from an established vineyard with a reputation for its flavor and ripening characteristics. They usually come from French vineyards, and one knows about them through reputable tastings.

I had been planting the rows at the top of the vineyard with the highest elevation. They will ripen first and be picked first as you pick the grapes for champagne before you would pick them for table wine.

By now, I had devised an approach to laying out the row for the vines with stakes and string. We would use the post hole digger to dig a whole about 20 inches deep and continue until the row was ready for planting. Fred would dig the holes for a row during the day and I would plant the vines when I got home from work. That spring we got the vines planted before the end of May. We were becoming seasoned in our work.

We had purchased a Hardi sprayer that attaches to the back of the John Deere tractor. It has lots of nozzles that need good cleaning between sprays. The theory on this type of sprayer is that you are putting a lot of liquid on the vines to ensure coverage, but a lot falls onto the ground (called *the vineyard floor*). We decide on a *sustainable* approach (in farming sustainable means working the land with low chemical input). This usually means more labor. Sulfur, an organic chemical, is a mainstay in the spray routines. A fungicide is needed in the East, and for the Vinifera plants since they are more susceptible to disease pressure.

The young vines were doing very well. So far, we were able to do all of the handwork on the vines ourselves:

- pruning
- gathering up the prunings and removal of the prunings for burning, which helps to minimize disease
- putting the shoots up into the wires to keep the canopy straight up (this is called *shoot tucking.*) Each vine has about 50 shoots, and each shoot has two grape clusters.
- leaf pulling to allow more sunlight to reach the grape clusters

The *canopy* is what one calls the whole vine of new growth. *Sunlight into Wine* is a book by famous Australian viticulturist Richard Smart and is a beautiful term for photosynthesis. A grape grower must always think about the sunlight to one's vines; that is the path to perfect grapes and wine.

Closer to harvest we are now having more "critter" pressure—groundhogs, raccoons and deer. A couple of deer can eat an entire row (240 feet) of grapes in one night! We started night patrols to scout the vineyard. My little puppy loves to help and chases the deer. We explored and tried all of the "hearsay" things to do like hang soap in the rows. We knew about various sound machines and tried putting a radio outside tuned to talk radio. The theory is that the critters would think it was people and stay away. That worked for a while until we found the deer standing next to the radio eating grapes!

During my life, I've had severe reactions to certain insect stings; and was especially careful since my sister had been rushed to the hospital with anaphylactic shock. One day, Fred asked me to get a tool that was under the tool shed. As I approached the tool shed, I was attacked by a bald-faced hornet. After running to the house for cover, I told Fred and he said, "Oh, I forgot to tell you about the hornet's nest." I donned my beekeeping protective gear to go back to eliminate the hornets' nest and discovered that the nest had been broken open—a sure way to agitate the hornets. I'm reminded of the press rod flying past my head missing my skull by an inch last year.

This year we would also harvest our Vidal grapes and try making an Ice Wine, which would be made by pressing frozen grapes. I didn't like sweetened anything but had tasted Ice Wine in New York that was delicious. The fruit sugars in wine grapes are fructose and glucose—no sucrose, unless added. I learned that I could drink wine grape juice without any problems. There must be a difference. I would understand much more as time went on.

I found a small grocer in Norristown that would rent space in his freezer for 25 lugs (a lug is a plastic picking bin that holds 40 lbs. of grapes) of Vidal grapes. The Vidal always produced a good crop with low disease pressure. We learned not to push the grapes, and to work with them each year to produce their best, not worrying about the yield. Quality was the only focus.

We transported the grapes to the freezer and waited months, till the temperature was about freezing outside, to press the frozen grapes. I measured the sugar levels of the juice (called *the must*). The measurement is in *Brix* (a specific gravity term of the amount of sugar in liquid).

It ended up at 36; normally wine *must* is 21-24 Brix. We monitored the press from sun-up to sundown, carrying a bucket of juice inside every few hours. After the must warms to room temperature, it's inoculated with wine yeast. This fermentation takes longer than table wine since the winery is getting colder because we don't have heat or AC, just the weather that nature gives us to work with. The wine yeasts eat the sugars in the must during fermentation and the alcohol level slowly increases. Eventually the yeasts get overwhelmed by the developing alcohol level in ice wine fermentations and the fermentation stops. At this point there is grape sugars "leftover" and that is the sweetness in our ice wines—but it is from the grape's fructose and glucose, not added sugar.

1996

This first Ice Wine was incredible. Our Vidal had tastes of peaches and apricots. By now, we had a few customers visit our converted milking shed on Saturdays and Sundays. The word travelled about the quality of this Ice Wine. I entered it in the Virginia Wine Association Eastern Wine Competition and was shocked to learn that we had won best of show–the coveted Monteith Trophy. I couldn't go to receive the award and sent a friend of the winery. He returned with the incredible solid silver trophy and stories about their shock that the trophy was going north of the Mason Dixon line, and of all places to Pennsylvania! For the next year, we had huffy visitors from Virginia demanding to see the trophy and take pictures of it. There was a constant effort by the Virginians to check on the Trophy–lest we sell it for silver value!

This year we were getting the 450 Pinot Noir plants that had been grafted for us. I staked out the first row and we got to work. Over the years, these vines proved to be very strong. We bought a used Club Car golf cart from an Amish shop in Lancaster County. It was gas powered and could handle a load up the hills. This was a big help in transporting vines in buckets of water, tools, wire, and us!

By now, we had many vintages of our young vines from Chardonnay, Pinot Noir, Gewurztraminer, Vidal, Merlot, Cabernet Sauvignon and Cabernet Franc. Each year, I used the Methode Champenoise process on both Chardonnay grapes and Pinot Noir, which are two of the classic champagne grapes. The Vidal would now be for the Ice Wine only.

All the winemakers I grew up with would do the same thing. If one used the Methode Champenoise process and the champagne grapes of Chardonnay, Pinot Noir, and/or Pinot Meunier, one called the result Champagne. It doesn't make sense to say Methode Champenoise when no one knows what that is. All of us revered French wines, French wine industry accomplishments, and French food. So, we chose not to put the word 'Champagne" on our labels, in deference to the French, but instead only Methode Champenoise, which is different than the soda-like CO_2 infusion. As I mentioned earlier, this is confusing by intent.

Years later the wine industry succumbed to the pressure that only French produced (from French vineyards in the Champagne district) Methode Champenoise could be called *Champagne.* On one of my visits to California I noticed that several California wineries put the word "Champagne" on their label. When I inquired about this, I was told that these wineries were now owned by a French Company, so they could call their wine 'Champagne' even though it was not made in France!

The cycle of work for a vineyard and winery follows the calendar year. January through April is heads down, pruning the vines. Then the growing season starts, and the vineyard takes daily care until harvest, which for us starts after Labor Day with the champagne grapes. Each of our varieties ripen at different times, so we harvest one variety and then on to the next through late October. The winery is heads down at harvest and continues to year end. The winter continues wine processing activities through bottling prior to the next harvest. Many vineyards grow one or two varieties, which will ripen at once and need a lot of resource at one time. Our vineyard has ten varieties, and lucky for us they all ripen one after the other; so, we go week by week through harvest.

Thus far the growing seasons are never the same. *Bud break* (when the vines start to grow in the spring) for us starts when the Chardonnay at the top of the vineyard (the warmest place) start growing their new little shoots. This is usually the last week in April.

The vine structure is comprised of a *trunk* (from the ground to the first wire) then the *canes* (last year's growth) are positioned and tied horizontally on the first wire. These canes have the buds produced last year, and they will produce the shoots of leaves and grapes this year. There's a saying about pruning vines that "you only get fruit from last years' wood."

One day I came home from work and Fred was sitting on the porch. This year had been cooler and wetter, and the fields needed more mowing as well as the vines tending. I didn't see the John Deere anywhere and held my tongue about why he was relaxing when there was work to do. I waited and finally he said there was a problem with the John Deere.

"What?" I asked.

"Brett borrowed our tractor because his was needing repair, was Fred's answer."

"And?" I said now getting concerned.

"He put gas instead of diesel in the tractor by mistake and the engine won't start."

"What!" I stammered. "How could that be? Isn't his tractor a diesel engine?"

"Yes, but he made a mistake."

I was incredulous. "How could you lend him our only tractor? They have two tractors, and another vineyard, and are owned by a well- financed couple. And they have several trucks with towing capacity and could have easily transported one of their other tractors. Why did you do this?" In the end, there wasn't a good answer. I turned my attentions to the how and when it would be fixed, and how we would have to make do until it was returned.

We had bought the used tractor from Yearsley in West Chester, a very reputable family serving Chester County farming from the beginning of time. They were gracious and offered the engine rebuild at parts and a fixed labor cost. The other vineyard owner agreed to pay the bill. We would pay the transport to and from. Ten days in the shop, if they could get all of the parts. I lay awake all night trying to figure out how we can proceed to get the work done. As a Project Manager in my day job, I knew to always have a backup plan, and if possible, a backup to the backup. The John Deere had become an indispensable team member–and we had no backup. Spraying the vineyard to prevent disease is paramount, since mowing can slide, if you can live with the tall grass.

We had completed a spray just before this happened, so we had about 8 days until the next spray. We worked the vines by hand as much as possible to put up the shoots and clear the weeds around the trunk. The grass was too high to try the push mower, so we focused on the *canopy* (all of this year's vine growth).

Our theory is always to do the best for the vine and the wine regardless of the amount of work. After this growing season, we decided to cut off the aging cordons and put a new shoot down on the trellis wire every year. This was a very wise decision since it removed a lot of disease that was harbored in these old canes, and also helped the vine to "self-balance" the fruit load. "Since you only get grapes from last year's wood," the new canes become the wood and have the healthiest buds and the right amount every year. More work, but it always paid off.

Finally, the John Deere was back. We took 2-hour shifts mowing through the weekend. The vines were ready for the next spray, with all of the extra hand work we had done waiting for the tractor to be returned. The mower had to be removed to put the sprayer on the 3-point hitch. So, we mowed around the clock, then changed over for the sprayer. It rained four inches the day after spraying the vineyard. The spray material lasts about 10 days unless it rains more than 2 inches per day. So, we sprayed again, and reset the 10-day spray clock.

This was late July, just before the grapes start turning color, called *veraison* (what a wonderful word!)–the sign that ripening has begun. We were starting to understand that wine grapes were intelligent plants. The vine will only take up a maximum of three inches of rain at one time. The plant shuts off the water uptake then. This ensures that the grape berries (individual grapes on the cluster are called 'berries') don't split their skins, as tomatoes would. This becomes more and more important as harvest approaches because two inches of rain will *decrease* the Brix (sugar level) in the grape berry by one degree. At harvest, we look for the Brix (sugar level in the grapes) to reach maximum. The grapes will slowly recover from the water uptake, but it takes about one week of sun to *recover* a loss of one Brix (sugar level).

More visitors were finding their way to our fledgling winery due to the good news about our awards. Over these first years, we had a signup sheet for a future newsletter.

This year we had our first Chardonnay champagne available. We had purchased a small winery champagne corker—referred to as "semi-automatic". This term is widely used in winery equipment, but really means mostly hand operated.

I remember our first weekend we opened for visitors. Karl, our carpenter friend made two wood signs, one for the entrance that I hand painted with our name, and one to be a guardrail sign which would be attached to the guardrail on RT 23 and read, "Winery Open" with an arrow pointing to Grove Road. I would drive up to the RT 23 intersection on Saturday morning and secure the sign to the guardrail with rope and take it off on Sunday evening. What a thrill it was to have a visitor taste our two wines and purchase something. The wines were priced at $10/bottle and two for $15. Our first weekend we sold about $45. That first batch of champagne was made with French champagne bottles; they have a different look, are heavier, and the neck is narrower.

After a couple of years, and after receiving several awards, we decided to have our first event for customers. We had the new champagne *vintage* (vintage is the year the grapes were grown) and our first produced labels. Before this, we had copier printed labels that we would cut on a paper cutter and paste by hand. There were about 125 customers on the clipboard, so we made up a postcard and sent it out. Our first Champagne Day was a complimentary taste of the new champagne and a sampling of hors d'oeuvres—brie, cantaloupe, and prosciutto. We thought maybe 25 customers might come and 75 did! They came dressed up and gathered around the old stone wall surrounding the paddock. Those customers bought all the champagne we had available for sale. We were starstruck that night and energized to keep going.

All year I had been travelling to Worchester, Massachusetts every week to manage a large enterprise project for Allmerica. The $40M project team of 40 consultants came from the U.S., Canada, Mexico, India, France, and South America. It took all my wits to keep the team moving. On the weekends, the vineyard and winery work were a welcome mental break. In December, I learned that I had received the IBM International Consulting Group Award and would receive the award in Madrid in February 1997.

1997

Fred would go with me to Spain. We planned to tour the countryside for three days before the IBM conference started in Madrid. We rented a car and took off for the many small villages and old estates. I had brought a wine shipping case inside a pull along luggage bag that would hold twelve bottles of wine. We sought out local wines and enjoyed small inns with restaurants. In Madrid, we found wine shops and bought a selection of Rioja wines to share with our wine tasting group.

In March, we started the cordon (these had the *bud spurs*) removal during the pruning months. We had been using hand Felco pruning shears and were surprised to find how hard it was to cut the four-year-old cordons. We had to carry big cutting shears; now we understood why they call it *wood*!

Also, I had discovered that our first Pinot Noir plants were actually the Gamay Beaujolais clone, from the region in France by the same name. It produced a lighter style than I was after. This is the grape used for the 'Beaujolais new wine" produced at harvest and released in about two months. I decided trying to make a red champagne the next harvest. It would be at least three years till we had the result.

The 1997 growing season started with the Chardonnay bud break on April 25th. The southeastern Pennsylvania spring frost-free date is May 10th and the date passed by with no late frost this year. Each week, the buds were popping from the top of the vineyard down the slope. The Cabernet Sauvignon and the Vidal were last for bud break, around May 29th. It's exciting to see this "coming alive" scene everyday—fresh green little shoots all along the lowest wire, about 30 inches above ground, decorate the vine rows. If you look closely, you can see the tiny shape of a grape cluster. Each shoot produces two clusters of grapes. It's easy to see why there are many different trellis systems, and approaches to get more grapes per vine. Over the next five years, we tried a few approaches but determined the vines didn't like them, for many reasons. After a decade, we were convinced that our vines did well on the vertical shoot positioned trellis. They were in good balance with our cane pruning work, and the disease pressure was significantly lessened. We didn't fertilize to push the crop, but did a soil amendment based on soil testing every five years. The vineyard wasn't a factory; it was a wonderful garden.

The first thing I learned about growers is that they want the grapes to be ready to harvest as soon as possible. This is understandable because the disease, insect, and critter pressures increase as time goes on. The best vineyards are south-facing, which means they get the most sun all day. That also means that grapes facing south will get more sun and be riper than the grapes on the north side of the canopy.

Veraison is when the sugar levels in the grapes start to increase, and the grapes turn color; then our attentions turn to protecting the grapes. Curiously, it is also when the water table starts to recede, and the vines push their roots down lower each year in the East. It's also a signal in nature to all the predators; they flock to the grapes.

In 1997, the growing season has been the best so far. Average rainfall is no more than once per ten days, with sunny days and a light breeze. The temperature in July ranged from a 90-degree high to 60-degree nights. The cool down at night allows the vine to rest. There was no excessive growth, and the vines are in very good condition for the last spray before netting.

We had purchased one roll of bird netting this year—10,000 feet, 17 feet wide, and cut it to fit the row lengths.

We would start netting the champagne grapes first, and when they were harvested, we would move the netting to other rows. Each vine row was about 240 feet long. We would pull the netting off the roll, measure out the length, pull the net down the vine row, pick up one side and 'throw' it over the 9 feet tall vines, then pull it down the other side. This is a big job, but it worked to keep the animals from eating the crop.

Champagne is considered the most technically difficult wine to make because of the many steps and the vast number of decisions created by those steps.

Methode Champensoise is picked on the lower Brix side because the wine undergoes a *second fermentation in the bottle* (this step is referred to as secondary fermentation) to produce the effervescence (bubbles), and that increases the alcohol about one degree in the finished product.

I knew how the tastes changed inside the grapes because I was tasting every variety, every week, after *Veraison* (color change) in our vineyard.

Many things are happening as the growing season progresses. Fruit flavors develop, color develops in the skins and seeds, and the acids decrease. These things are watched closely as harvest approaches and then one day the best ripening has happened, and the race is on to get the grapes picked and pressed.

By now I was very aware of many important considerations in the winemaking:

- Temperature at pressing
- How long to settle the newly pressed juice before *racking* (pumping) to a clean container
- What yeast to use
- What nutrients are needed
- To add sulfur at press or not
- Temperature of fermentation

And the list goes on since each variety is unique.

This harvest, I experiment with two different yeasts for the Chardonnay that will become champagne. During the champagne making process, the pressed juice is initially fermented to make wine, and then a second fermentation is induced to create the bubbles (*effervescence is the term used to describe these bubbles*) in the champagne bottle. I discover that using a different yeast for this second fermentation can add a slight enhancement to the overall flavor.

The outside temperature during harvest this year was not overly hot. The white wines are best if pressed when it is cooler, and fermented in cooler conditions, but they are first to harvest when the weather is hotter. The reds are the opposite; best to press and ferment when the weather is hotter to get more color from the skins, but the reds are harvested last when the season is ending and the temperature is falling.

1998

The winter was mild and our cane pruning was going well. We began working on the lower area of the dairy barn to rehabilitate it for future winery space. It took me two years to repair the grout and parge coat the stone walls with white masonry cement.

That spring we continued working to finish painting the house outside. One second story area was accessed with a ladder on unsteady ground. Fred was holding the ladder, as I worked on a vent window. For some reason, Fred left the ladder unattended, and it started to fall. I was able to grab a window ledge and break my fall. I made a mental note not to forget this dangerous incident.

At IBM I was now working in the Mid-Atlantic region with small and medium sized businesses. I travelled that winter to Williamsport for a project at Brodart Company, who specialized in schoolbooks. Over the summer I worked with Medco in northern New Jersey who was pioneering a customer care system to supply prescription drugs directly to customers.

I started a red improvement program this year, since I knew our red wines weren't that good. I was confident in our white wine results, but the red wines required long aging to "show their strength." I wasn't sure whether the grapes or the winemaker was not good enough.

We invested in 100% French Oak barrels at $500 each instead of American Oak barrels at $200 and experimented with adding tannins at fermentation and new Bordeaux yeasts from France. The colors at pressing were darker; we were on the right path.

1999

We continued working on the buildings. This year our focus was on repointing the old barn, which meant chipping off the one-inch concrete, sand blasting the stone, acid washing and finally repointing. Inside the old dairy barn, Rolland was excavating the floor down one foot and reinforcing the stone walls around the interior walls. These old stone barns were laid directly on dirt, without a foundation, so we were very careful and exposed only three running feet at a time, as we dug down one foot. Then we would 'pour' three feet of foundation and move to the next three-foot section, until we had completed all four walls in the bottom of the barn. It took about a year to complete this foundation work, then we poured a concrete floor with drains, thus adding one foot to the height of the room.

Five hundred seventy-five new Cabernet Sauvignon vines were planted in early May. From May 30th to August 15th there was no rain in the East. This drought destroyed many crops. In June, we filled buckets of water, and used the golf cart to take water to each of the newly planted vines. It took five days to take one gallon to each vine. Then we used the 80-gallon sprayer tank on the tractor to deliver water to the new vines. As the weeks went by, we constantly brought water to the new vines and worried about the older vines, since Lancaster County farmers reported one crop failure after another.

In the East, there is usually plenty of rainfall throughout the year for crops, hence no need for irrigation. This was the first year known in the mid-Atlantic with a ten-week drought.

We applied for a new well and were approved in mid-August since we were in agriculture. The day after we got the approval notice it rained.

Shortly after my 49th birthday that summer, IBM changed the retirement plan. Overnight, I lost half of my retirement benefits. When the shock cleared, I started looking for a new employer. I had helped my eldest sister, who owned a regional staffing company, sell an SAP data conversion project and in 2000 joined the company with a 10% equity promise.

2000

With the continuing outside income, we decide to go ahead with winery expansion plans, including a spacious tasting room for visitors, and new wine production and storage spaces.

We start repointing the outside of the barn. Scaffolding was set up to remove the inch of concrete, then acid washing the stone, and new pointing. The work was slow and took the year to complete.

The driveway was packed dirt with some crushed stones. We decide to put new 2B stone the entire length of three hundred feet. We started in the dry time of summer; one truck a day was dumped, and we spread the stone out with big rakes. Finally, we rented a tamper to pack the stone into a firm surface.

We decide also to put in the new well this year, because we will need it for the winery expansion. We'll also add capability to irrigate in case that is needed again. The older vines all came through the drought with no problems, giving us a deeper understanding of how unique wine grapes are. We knew that each year when the water table recedes in the east, the vines push their roots a little deeper, seeking water and nutrients. This helped them withstand the drought. I'm starting to understand why at the Virginia Tech Viticulturist seminar, one researcher described them as "intelligent plants."

Soon, Botrytis (one of the four major wine grape diseases) breaks out in the Chardonnay, so the champagne crop is at risk. Daily, I scout for infected clusters and cut them out, collecting the clusters in a pail to remove from the vineyard. Later, Fred says he missed a scheduled spray.

At my day job, the small company had grown dramatically from the Y2K work. This was the computer problem as some computer programming didn't go past 1999, and the fear was that the programs would not work in the year 2000 (Year Two Thousand or abbreviated to Y2K). I discovered that the new company had no backlog of work for the consultants and start searching for new work. I find a project that I can do myself as a consultant; anything to bring in revenue.

All the grape varieties are bearing a full crop. It takes three years to get a partial crop from wine grapes and five years for a full crop. For the white wines, we have Chardonnay (both for table wine and champagne),

Gewurztraminer, Viognier, and Vidal for Ice Wine. For the red wines, we have Cabernet Sauvignon, Cabernet Franc, Merlot, Petite Verdot, and Pinot Noir. The first four red varieties are Bordeaux grapes, and these varieties can be combined to make what is called a *Bordeaux red wine*.

This year would be my first attempt at a classic Red Bordeaux wine in the image of a French Red Bordeaux wine. I acquired four different French Oak barrels for the four red Bordeaux varieties. Each variety ripens at different times with Merlot first, Cabernet Franc second, Cabernet Sauvignon third, and Petite Verdot last. It will be extra work to keep them fermenting separately in these small quantities, but important to have the barrels with only one variety, so when time comes for blending trials, we'll have an accurate picture.

This year our Champagne Day hosts almost 200 visitors to taste the new champagnes. The 1997 Blanc de Blancs was the first *vintage* (the year the grapes were grown) from the top of the vineyard planting and was a small yield. This day we sold half of the vintage; that was all we had finished at the time. There were four couples that purchased for their weddings, including Cynthia Baughman who later worked on the U.S. Air Attache Magazine article. Everyone seemed to know that this was a "cut above" anything we had ever produced.

In December the new Pennsylvania state viticulturist, Mark Chen, recently hired from Oregon, encouraged us to enter the Vinalies Internationales wine competition in Paris, France. This was thought to be the Olympics of wine competitions.

We knew that wine judging in general was very hit or miss and dependent on the "preferences" of the judges. Mark explained about the *reproducibility factor* for the Vinalies judges. *Reproducibility* in wine judging is established this way: A potential judge goes to a specific site that is set up with individual judging booths. They are given a flight of wines to judge each day for 5 days. Some wines are duplicated in the flights. Their analysis is compared. If the same wine was not judged identically; they don't pass.

The Vinalies competition requires the judges to pass this reproducibility test.

We investigate entering the competition. You must send six bottles, and the entry fee is $600. It turns out that all the entries from North America at that time were from the West Coast and the only shipping was by container from LA. We would have to air ship the wine to a carrier drop in CA.

We explore finding a customer or relative going to Europe, and in the end decide to take it ourselves on the plane. Champagne is delicate and lots of shipping transfers can 'wear' the wine. We send the entry form and payment in December and plan our trip for early January, as the samples must be in Reims by January 20th. I find a translator who will join us for a few days as we plan to visit a research vineyard and Champagne houses. I have been researching champagne making equipment for small operations. Methode Champenoise is a many-step process and each step requires different equipment. The steps are spread out over many years; there is no bottling line solution as in *table* wine (*table* wine is non-sparkling wine.)

The overall steps in making Methode Champenoise include:

1. Grow the champagne grapes (Chardonnay, Pinot Noir, Pinot Meunier)
2. Pick for champagne parameters (lower *Brix* or sugar level, and higher acid)
3. Fast pressing and settling at low temperature to avoid off flavors
4. Low sulfite additions—as that may prevent the future second fermentation
5. Continue from harvest the wine making process with winter cold stabilization and racking (pumping) into clean containers

6. In the Spring begin preparations for the second fermentation—purchasing champagne bottles, yeast and crown caps (same as beer caps)
7. Grow the yeast starter, filter the base wine, then add nutrients and the yeast starter. (requires a filter system, and laboratory equipment for the yeast starter)
8. Bottle into the champagne bottles (bottle filler, pumps, and crown capper machine)
9. Store horizontally in secure *boxes or metal cages* for 18 months to 10 years.
10. When ready move the bottles to a *riddling* device. (This machine slowly moves the sediment in the bottle to the neck in preparation for removing the sediment prior to final corking)
11. Move from the *riddling* device to containers in the upside-down position (to keep the sediment in the neck of the bottle)
12. Move containers to the next processing area to rest for a time.
13. The final processing steps are to:

    a. *Disgorge the sediment* (remove the sediment via a special machine that quickly opens the crown cap and the sediment is released with a small amount of the wine)

b. Apply the *dosage* (pronounced *do-saaaaaajjje*). Dosage is the liquid that is a "House" recipe, and includes the sugar amount desired, which is referred to as *Residual Sugar or RS*.) More information below as this determines the classification of Natural, Brut, Dry, Sec, etc., that is always listed on the label.

c. And finally, insert a champagne cork and apply the wire hood. (Requires a special corker working under pressure and a wire hood machine)

Many years later we made a 14-minute video of this process showing all of the equipment and uploaded it to YouTube:

https://youtu.be/7ULpeVFAPjw

Dosage Classifications of Champagne (percent of residual sugar)

As defined by the Oenologues of France organization

Natural	0
Brut	0–1.5
Extra Dry	1.2–2.0
Sec	2.0–4.0
Demi Sec	4.0–6.0
Doux	6.0 ~

So, what does this mean? The "driest" champagnes are classified as Natural. These are generally scarce. The most commonly consumed are in the Brut classification. However, notice that the range of sugar for Brut is quite large. That means you could get a Brut as "dry" as a Natural—with 0% residual sugar—or what I would call sweet at 1.5% residual sugar. That's pretty confusing. Thank our French friends for that.

All my champagnes are finished with no more than

0.5% residual sugar. That is why they all taste very dry. I prefer dry and think these wines and champagnes pair better with foods. That's my personal opinion.

2001

In January of 2001, our trip is fast approaching. The Vinalies entry has been accepted. We just have to get the champagne to the competition headquarters in Reims, France. Our first stop is Epernay where we meet the translator. Sunday has most restaurants closed, and we are told the train station has a small restaurant. We are pleasantly surprised that we can get a glass of local champagne with our sandwiches—that's the intent behind *Vin de Pays* (a technical French classification term, but in everyday language means local or regional wine).

The next day we head out to a research vineyard. We're surprised to see the same trellising we use—vertical shoot positioning (called VSP), and the same spacing we had in our vineyard (4ft by 8ft). We ask about all the different trellis systems and spacing which are currently the "rage" and they reply that after a century of analysis, this was the best.

Next, we visit a champagne maker's house to see their

cellar, equipment, vineyards, and taste the champagnes. We discussed every step of the Methode Champenoise process and the equipment they used. They explain why they make *Charmat* in some years. Charmat is 2 atmospheres pressure - hence less bubbles, and much easier to make. This is a good solution if the grapes don't ripen to the best parameters.

Methode Champensoise is made between 4 and 6 atmospheres pressure, with 6 atmospheres giving the most bubbles, but harder to handle along the way. This pressure is why there is a punt (depression) in the bottom of the champagne bottle, and the glass is much thicker.

Their cellar predated WWII and had a two-story elevator below ground. We were told the story of why the Germans so wanted to claim this area of France. They coveted the Champagne cellars with thousands of bottles underground!

We brought the Vinalies entry bottles packaged for shipping in France, as the competition headquarters in Reims wasn't a destination for our trip. The translator offered to hand deliver the wines, as he worked down the street. We handed the package over trusting that they would arrive at the competition headquarters.

The next day we visited Oeno Concept to look at an automated riddling machine called a Gyroflex. We had seen a palette sized operation needing a warehouse, and heard they were working on something for a small operation.

Riddling is moving the sediment down the bottle into the neck of the bottle very near the top. We had tried various American ingenuity methods such as:

- Boxing 12 bottles and giving them a little shake every day.
- Driving the boxes around in our van for a couple of weeks
- Loading the tractor cart with boxes and driving them around the vineyard

Unfortunately, all three methods had limited results.

We were ushered into the general manager's office for a sales presentation, thinking we were wealthy Americans. After explaining three times that we really were very small, and without a warehouse, we thanked the executive and turned to leave. Halfway to the door I hear "Mademoiselle un moment" and turned around.

For the next half hour, we were shown pictures of the new entry level Gyroflex that would handle 150 bottles, which was perfect for us.

The executive was interrupted with an international

call. We heard him explain that the equipment was delivered as promised in China. Many questions later, he got visibly annoyed with the caller and said, "We will not come there and make champagne for you. You don't even have any vineyards yet."

We left with instructions about the various champagne bottle molds fitting the machine, and to send samples of our bottles, so they could have the machine customized to hold the bottles with a snug fit. At home, we would need to arrange shipment and customs clearance, but it would be a six-month wait.

We had one more equipment visit to TDD Grilliat in Epernay. They made equipment for the very tedious and time-consuming finishing Methode Champenoise process we refer to as "disgorging." It involves removing the crown cap, topping off the liquid, adding a dosage, corking and wire hooding. These steps require picking up the 3½ lb. bottle 10 times! In one session, I would work on 12 to 20 cases; that's 144 to 240 bottles. A twelve-case session would require moving over 5000 pounds (two-and-a-half tons!).

The TDD Grilliat equipment could replace many of those hand steps with automation. The machine we were most interested in cost $75,000. We hoped to find some older refurbished and more affordable equipment that would eliminate some of the hand work. We went home with lots of brochures.

We finished our delightful stay in Epernay with breakfast at our hotel bar, which was the usual. This time the owner gave me a champagne stopper (as I call it) to take home. This is a simple device to hold in the effervescence after the cork is popped. I had asked about it when we first arrived. It is still today the best available, and a treasured gift.

At home, we put the competition out of our minds since it would be months for the results. The Vinalies receives over 4000 entries from around the world, and they only award Gold and Silver. The winner of each round of judging advances, and a bottle is used in each round. The six-bottle entry is used if you make it to the final round. The judging takes about two months to complete.

We busied ourselves with pruning activities. One day in late March, we received a message from the new Pennsylvania state Viticulturist, Mark Chen. He got a call from a wine judge he knew in France to say that we had won the Medal D'Or (gold medal) for our 1997 Blanc de Blancs. The judge wanted to know where *Pennsylvania* was!

We were totally surprised and cried.

The growing season starts right on time with the

Chardonnay buds popping on April 25. The barn swallows arrive on the same day. We await anxiously to receive our gold medal.

A letter arrives from France officially notifying us of the results.

I email the competition office and ask when will we receive the gold medal? They reply that we won't get a medal–they give out diplomas. I immediately think of the blank diploma forms at Staples used to print participant's names.

I travel two to three days per week in my day job with Information Technology (IT) project consultants to San Antonio, Houston, Chicago, New Jersey, Philadelphia and Washington, D.C. This allows me only two weekday evenings for vineyard and winery work.

We decided to create our first winery website and offer online sales and shipping for our customers. With the newer vintages, we worked on better label graphics with a local company. Keeping costs down is always paramount, so we stayed with our hand operated label gluer. On the weekends, we offered tasting and sales in our little milking shed room with a kerosene heater where we wash, label and capsule bottles for sale at the same time.

Outside, we finish the pointing of the barn wall and prepare to break ground for our new addition to the barn. The old cistern will have to be removed and new electric service installed. We discover that our electric costs will increase as this new service will be considered "commercial."

One day in May, a special delivery arrives—a large tube. Inside we discover the Vinalies "diploma". It is a work of art—hand calligraphy and gold leaf! What a treasure this is!

The 2001 growing season is perfect weather with sun and limited rain. This harvest we will finally have a poured concrete press pad on the backside of the barn, instead of the makeshift plywood area over stones on the side of the house.

We break ground in June on the winery addition to the barn. It will include a spacious visitor tasting room, upstairs office, basement, and winery working space which will allow us to grow to 2500 gallons, or about 1000 cases.

In July, I meet with my eldest sister's company accountant to discuss taxes and the consulting business financials. He shows me the three-year projections. I notice that my 10 % company position is not included. When questioned, he replies, "Your sister has made no allowance for that. She believes your offer letter would not stand up in court.," and added, "I thought you should know."

I start looking again for another job. After Y2K, the IT consulting business is falling apart. Companies have spent more than planned on new systems and the country is flooded with H1B visa consultants, who will work for lower wages than Americans just to stay in the country.

We send out a winery newsletter announcing our Vinalies award and enjoy some regional press and customers.

By now, we have mastered the pre-harvest vine preparations of trimming, spraying and bird netting. Night patrols help keep the deer at bay. This year, I put up 20 bird scare balloons in the vineyard. It helps and looks festive for visitors.

I meet with the Pennsylvania Department of Transportation (PennDOT) regional management about getting winery tourism signs installed. New York had created a tourism signage program that had national approval, but Pennsylvania had yet to implement. After some pulling and tugging, they reluctantly agreed for signage at the RT 23 turn onto Grove Road. This national winery signage was a blue and white sign with a bunch of grapes and a turn arrow. The township had to approve and after more pulling and tugging, they did. This meant I didn't have to tie a handmade open sign to the guardrail anymore!

On September 11th, I arrived at the office parking area on a conference call and remain in my car with the radio tuned to New York sports. A little before 9AM, the unthinkable is broadcast that the World Trade Towers have been struck by an airplane. I run into the building to see the coverage on TV. Shock is the only awareness. For the next three weeks, there is not one visitor to the winery.

Champagne Day 2001 in late October is attended by over 200 again and the new building addition is well underway. We introduce our first J. Maki label on the 1999 Blanc de Blancs with its distinctive gold with black lettering and it's a hit.

Finally, the new addition windows and door are installed, and we plan to open the new tasting room in December.

Harvest finishes late, nearly Thanksgiving, with gorgeous warm, dry weather. The red wine *berries* are small with intense looking color. The small berries mean less volume. We don't care because it also means higher quality.

Our new website offering shipping, and the Vinalies award yields increased holiday sales.

The ice wine pressing goes smoothly—ten days of pressing frozen grapes, ten gallons per day yield. Inside the winery, the fermentation goes slowly in 65 degrees. Because of the high sugar content, the yeasts work slowly to eat the sugars and ferment them into alcohol. The aroma of the fermentation this year is especially aromatic with fruit flavors, the result of the great growing season.

2002

In January 2002 we begin to think about a bigger automated press and research the used winery equipment at the next industry show. A $20,000 used press could process three times the volume of our one-ton basket presses in one-third of the time, and with half the cleanup time. We decide to visit the equipment broker for a demo.

Later, we order the used press. It will arrive from Germany and be reconditioned for delivery prior to harvest. We'll have to pay the overseas shipping, transportation from Virginia, and the customs duty. Once on site, we'll have to figure out the electrical connections.

Winter weather is mild, and we make good progress in pruning. I have taken over pruning the Vidal, the Gewurztraminer, and the old Cabernet Sauvignon. I have more life-time plant experience than Fred and can more easily make the hard "has to be cut" pruning decisions, it turns out. As the vines get older, they grow the way they want to—not the way you think they should!

After the Paris award in 2001, we see many new customers and visitors from far and wide. A few visitors ask, "How can you call it Champagne?" We explain about the process of Methode Champenoise and that winemakers always call the resulting wine *champagne*. Later, we explain about the difference between Methode Champenoise produced champagne and Sparkling Wine which is made like soda, with pumped in CO2. Later, we add that there is no patent on the word *champagne*. And still later, we add that *champagne* is an English word and was brought to France by the Lord of Champagne. And finally, we say that the French "*champagne* police" have told everyone not to use the word, unless you are French, and the wine writers are going along with this.

I decided to perform soil tests and make "amendments" based on the results. Lime is always needed, and a fertilizer mix based on the results. We use our old Dodge Caravan to haul the lime and fertilizer from Lancaster County. I filled five-gallon buckets, one for lime and one for the fertilizer, and set them on the golf cart floor. We would drive up and down each row stopping at each plant to pour the mixes at the base of each trunk.

Bud break for the top Chardonnay is again the 25th of April. Both the hummingbirds and the barn swallows arrive the next day.

By May 1st, the other white grape varieties, and the Pinot Noir are pruned and tied down, ready for *bud break*. We are always pressed for time in finishing pruning. The Cabernet Sauvignon are the last to start new growth, so we finish pruning those grapes in the early days of May.

Bud break is the most beautiful time in the vineyard when it comes alive with tiny verdant shoots bursting out of the buds. It's amazing how fast the shoots develop to fill the trellis. By the end of June, the vineyard has full foliage growth and the vines then work to grow the grapes. By August 1st, the green grapes are fully grown and turning color. This cycle of life culminates with harvest every year.

At the end of July, we decide to start netting the Gewurztraminer early this year. They're at one end of the vineyard, and last year the critters started eating the grapes around the ends of the rows early. The bird netting is packed on wooden pallets and covered with tarps for storage outside. After unwrapping one pallet, I load the Gewurztraminer nets into the golf cart. I roll three nets out in the rows, and call Fred to help pull them over the trellis. After five minutes, Fred is gasping for breath, and leans over as if to fall to the ground. He breathes heavily for five more minutes before calming down. I get him in the golf cart to rest and sip some water. He says he's okay and is just a little light-headed. Yesterday was very hot, and he worked outside for many hours. I suggested that he go to our local doctor after the weekend. For the rest of the weekend, I work by myself to get the three rows of netting installed.

The summer has been steady with new customers and visitors to taste our famous champagne. Everyone is happy to talk with us and wish us well. It gives us energy to keep going.

The netting goes up slowly. We only have nets for about half of the vines. After September, when the white grapes are harvested, we can move the nets to cover the red grapes.

By mid-August, we have done as much as we can to protect the grapes and are testing weekly to monitor the ripening. We turn our attentions to bottling as much as possible to free up the containers in the winery for this year's harvest. Of special note is the 2001 Vidal Ice Wine which has the most explosive fruit *expression* of peaches and apricots so far (*expression* is a term used to signify words that define the aromas or flavors in the resulting wine). Customers would sometimes think it meant we used peaches and apricots to make the wine, but it means to describe the aromas and flavors from the wine grapes.

One day in August, I'm in Philadelphia meeting with the

President of United Way discussing working with them on a planning project. Traffic is jammed up on the way home. There's a message from Fred that he's on his way to Reading Hospital and will be home later. Alarms go off in my head. He had gone to the doctor last week but said it was nothing. I call the doctor's office and find out that his red blood cell count was dangerously low, and he was sent to the ER for a blood transfusion. More tests are recommended.

The last days of August are the "calm before the storm" of harvest, which is always non-stop, 24 by 7 work. The Chardonnay for champagne is always the first to pick and it will be by Labor Day this year. My *Brix* (sugar level) target for the Blanc de Blancs is the same as the gold medal vintage: 19.6 Brix at harvest. For our vineyard, the fruit flavors are very delicate in that range, although no two vintages are exactly alike.

People often ask why we have a date on the label. We explain that is the year the grapes were grown. We work to get the best flavors each year. There are two approaches regarding the growing vintage. First, do the best with each growing season, and the date is on the label. Second, blend the vintages to produce a certain "profile" and the label is not dated. Had we taken the "profile" approach the Gold Medal vintage would not have existed.

After the champagne grapes were picked and pressed, Fred had additional tests and we are told he has Stage 4 Esophageal cancer. The local doctors tell us best to get our house in order. We decided not to tell anyone.

When the shock from Fred's diagnosis clears, we seek out another opinion. Through my Pastor Cindy, a friend at the YMCA, we go to Fox Chase Cancer Center. They say they think they can help us. Fred is big and strong, so we decide to start treatment immediately. It will be every day this fall, and then surgery.

The growing season has been very hot and dry. The reds are pressed late, some people call it "hang time" in the vineyard and may be very good. The berries are small, low yield but more intense flavors.

I decide to start pruning in November instead of January, since Fred won't be able to participate. I drive to Reading every day to pick up our two workers and develop a technique I call "pre-pruning". We'll cut out all the old wood now and leave the final cuts (the decision cuts on what will grow next year's grapes) till the last minute in the spring. The vines don't wait–you must have the work completed.

Fred is having chemo and radiation, and steroids to help eating. He has become over- active and wanting to purchase things. This morning, he said we had to go to Lancaster to sign the papers for a new tractor.

The winery has been closed many days since August, because we are at the hospital for Fred's treatments. We are slowly losing sales. Wineries are retail businesses and expect 50 % of annual sales in fourth quarter and 25% of annual sales in December.

Our December is slower than last year. Customers complain that they came here this fall and we were closed. We were at the hospital many days but just say "We're sorry—we couldn't be here that day." Our published information and phone message says our hours are "Saturday and Sunday, Noon to 5PM," and weekdays "as available." Fred's appearance belies his physical state and we carry on through year end afraid to speak about the upcoming surgery in early January. Only my family knows about this.

I completed the engagement at United Way in December. It was very challenging, juggling everything with a little help from family and friends. At home, Fred has papers spread out before him and wants to talk. He suggests we enter the 2001 Vidal Ice Wine in the Vinalies Competition. It is by far the most outstanding vintage and everyone who tastes it is blown away with the very delicate fruit flavors of peaches and apricots, and a creamy light sweetness.

"But" I say, "the French (The Vinalies is a French competition) hate the Germans, who invented Ice Wine (*Eiswein* in German)."

"Yes," he says, "but we know from our customers who visit France and take our Ice Wine as gifts, that the French do like it."

It would be good to have a second Vinalies award; no one in North America has two!

We decide to enter and prepare the entry form. There is international shipment available from the East Coast now through Fed Ex at the airport. We carefully pack the ice wine bottles for air shipment and put it out of our minds as the competition results won't be known till late March.

2003

In early January, Fred undergoes esophagectomy surgery, which is removal of the esophagus and reconfiguring the remaining stomach to attach to the lower throat. The surgeon is happy with the radiation results. We live in limbo for two weeks as Fred moves from ICU to a patient room and stabilizes to go home.

Big snows occurred up to three feet this year. We must wait for snow melt to get into the vineyard.

Cynthia Baughman is working with us on a U.S. Air In-Flight Attaché Magazine article about the win in Paris. It publishes in February, and we are blessed with many new customers. She and her husband had purchased the gold medal champagne when it was released before the competition, and they had it for their wedding. Her husband is a doctor and was treating Tug McGraw, the major league pitcher. They visited the winery last December and somehow, he instinctively knew that Fred was sick, even though we had told no one.

They came back to visit again, and he brought Fred a good luck coin.

The thirty-inch snow mass in the vineyard is hard to deal with, but you can only wait so long to prune. On February first, I punch down a path up to the top of the vineyard where the snow will melt first. Dressed in a mountain climbing Gore-Tex wind suit, snow boots, and wearing my 25-pound battery operated pruning shears, I trek up every day possible to make progress on pruning.

Fred is recovering from surgery but has difficulty eating. Most of his calories are from a feeding tube and some soft foods.

The snow is finally receding in March. The driveway and roads have had ice every day. I don't know how I will ever complete the pruning and clearing the old wood from the vine rows, but I just keep going. In late March, we receive a special delivery letter from France; our ice wine was awarded a silver medal—Medal D'Argent!

May—We built the winery little by little. I don't know what possessed me to believe that we could be successful. Everyone in the wine industry knows that small wineries don't make any money; even with winning a gold medal in Paris. It's a lifestyle we say. Some things are changing, but we're in Pennsylvania. If we were in California, that accomplishment would have created a tremendous stir.

We can't even get the regional press to cover it! Secretly, I have always hoped and planned that it would support us. But the sad truth is that it doesn't yet and won't for many years. I have invested my life's funds. Why was I so sure we could do this against all odds?

May 19—Winery Spring Fling Event. I decided to try another customer event and give tours of the vineyard as the new shoots are popping. It was a big success. Customers were excited to see the little grape clusters on the new shoots.

May 22—Vet appointment day. Our dog Tashie is jaundiced. They say she has a serious illness, possibly two things. I breakdown crying immediately, apologize and say my husband has cancer and it's been a tough year. They're waiting for other test results, but Tashie should get an ultrasound asap.

The specialist performs a very sophisticated bypass for the pancreas and the bile duct the next day and thinks there is no cancer. The next week we get the test results. Tashie has pancreatic cancer.

June 2003—In early June, I travel Friday, Saturday and Sunday to the wine festival in Allentown.

I poured about 1000 tastes per day, but sales were minimal. Most of the attendees are there for a good time, and to get drunk. I hate this about wine festivals in this state.

Mostly, you see arms held out in front of you, sticking their glass at you saying, "Gimme some of that!" This was a Breast Cancer Benefit.

During the week, I talk to Fred's Oncologist about veterinary treatments for cancer He advises that chemo alone will not stop pancreatic cancer. I ask would he treat if this was his dog? No, was the reply.

We're busy in the winery with more USAIR article customers; thank goodness for airline flyers that read the magazine in front of them.

Tuesday, June 17–It's still raining, 40 days and 40 nights. Amazingly, the vines seem fine. The work to convert the vineyard to *cane pruned*, is paying off big; otherwise, we'd have a jungle of shoot growth. Now we're just about 10 days behind in getting the shoots up. But we need sun, and no rain, for about a week.

Sunday, June 22–The rain has finally stopped.

Monday, June 23–The sun is out and it's a beautiful day. A traveling salesman stops to see if we want our old barn roof painted. We were planning to do something this year. We discuss and look at pictures, ask for references, and in the end, say yes. We believe there are still honest people. They arrive in 30 minutes and start the power washing.

Tuesday, June 25–This is a very hot day. The *leaf botrytis* (one of the four major grape diseases, and prevalent if the leaves are wet) is surely gone. I go out early for errands, and to see the graphics folks about our T-shirts, new labels, the custom label for Drexel and the PLCB bar codes.

Five days after six inches of rain, the ground is getting hard already. That's the benefit of south facing slope and well drained soils. There's still time for the vintage to be good.

Fred has a follow-up with his doctors at Fox Chase. He looks good and they say, "See you next year!"

July 6–This is another hot, humid, typical July day. Some of the Vidal has powdery mildew (one of the four major grape diseases).

I struggle with the Pinot Noir. It's overgrown, more than ever before, and there are shoots under the graft buried in dirt. It takes two hours to clear eight plants. I'll never get the last three rows done anytime soon. I ask Fred about the *grape hoe* (a tractor attachment that pushes the dirt up around the graft area of the vine to protect it from winter injury) on the hill. This technique was widely thought a must in the Eastern U. S. wine grape growing regions, but over time we learned that it wasn't necessary and abandoned it. I can't operate the attachment therefore I must clear the dirt by hand–400 plants in between the other vineyard work.

July 7–Our one vineyard worker arrives upset and barely

able to talk. His father died, and will we give him money to go home to Mexico? We do and he says he will return soon. His brother will work until he returns, but we have to transport him to and from Reading each day.

July 9 – The vineyard is in a crisis with 80% more growth than last year. All hands work all day and get seven rows of Gewurztraminer done, only 120 more rows! Fred is only able to help for the morning. We sort out the shoots, clipping off the extra non-fruiting canes, and clear the new growth around each trunk, all the way to the ground. Then we begin the slow work of stuffing up all the shoots, one at a time, up through the three pairs of wires.

Friday, I get the bad news that I didn't get the last job I interviewed for.

July 22 – Yesterday the Pennsylvania Premium Wine Group (PPWG) meeting was here, Chaddsford Winery is our newest member. The group has come a long way fast, by-laws are approved, officers elected, and technical recommendations approved. We're running fast to the first Pennsylvania Premium Wine Tasting Panel, lots of good publicity to happen; should be exciting for the rest of this year.

July 29 – Today is my 53rd birthday. Fred is not feeling well.

August 1 – We disgorge champagne. Fred is trying to help

out as much as he can. The new corker stops in mid cork. Four hours of fiddling and calls to the technician, I end up taking the jaws apart and cleaning them very carefully, since they are almost as sharp as the microtome knives I used in the otology lab. Now I know how to do that and will plan on maintenance.

Our friendly Mennonite electrician arrives later that day, and he and I run through the things we learned last time about the riddling rack. It's three phase and he had to install a 3-phase converter in the basement. After five minutes of checking fuses, relays and currents, the machine starts working.

August 17 – Veraison (color change in the grapes, increasing sugar levels) is finally happening, a signal of ripening. Pressure is on for the final prep before the bird netting goes up. We trim, leaf pull, clean under the vine rows, and spray for botrytis. This is a costly grape growing year in time and money, and the results are in question. The winemaker earns her keep this year.

August 19 – We will sell the new press because we need money.

We hear about the short article Marnie Old did in Philadelphia Magazine about local wines. She gives us a very nice mention, talks up the Gold Medal Blanc de Blancs, and reviews our '99 Cabernet Sauvignon.

Saturday, August 30—The crummy weather continues, with scattered showers keeping everything wet in the vineyard, and the disease pressure up. The grapes are slow to turn color. We net the Pinot Noir two weeks earlier than usual, since the insects, birds, and deer are all over it. We need the weather to be dry, and the summer to last through October.

We sold the press to a Washington State winery.

Friday, September 5—We need to get the Gewurztraminer netted, since the varmints have been at it. The deer are also hanging out in the Chardonnay in broad daylight! Tashie is working hard now, checking the vineyard for deer, raccoons, groundhogs, and birds.

Sunday, September 7—Yesterday we held a customer event with vineyard and winery tours. We're mobbed with people, best single sales day ever. Mom is the greeter and Dad shows up around 2PM. He goes on the tour and keeps watch out front, slowly pacing with an Indian Chief glow. He's bruised from Coumadin; 82 years old, and still going. My parents constantly inspire me.

Tuesday, September 17—We go back to Fox Chase for a CT scan. Fred is not feeling well. Some spots showed earlier in the year and now there are more. The doctor calls for more tests before treatment options.

Thursday, September 18—Paul arrives to help finish the

courtyard landscaping. We haven't made much progress since he was last here in the spring. I make runs for sod and edging stone. By 11AM, we are underway, and making progress. Miguel and I lay sod and Paul works on the edge. It's a race before hurricane Isabel arrives later in the afternoon. At the same time, we are picking Chardonnay for champagne, and get over a half ton processed before the rain starts.

Hurricane Isabel rages on through Friday. We must get the Chardonnay for champagne picked. On Saturday, we have a five-person harvest crew to pick and press as much as we can.

Champagne Day, October 18 – The last two weeks are a torrent of work and activities to do with harvest; bottling reds, barrel routine, and getting ready for Champagne Day. We are ready for the 400+ guests.

November 10 – Harvest continues but is finally winding down. We took three loads of Vidal to the freezer for our ice wine, which was half the normal crop. This year, we did three times the work, spent twice the money, and will get half the crop. Tomorrow, the last Cab Sauvignon grapes will be picked. Pruning will start tomorrow or when it isn't raining, the story of this year.

December – We're looking for close to 32-degree F temperature so we can start the ice wine pressing.

Where are the customers? The room is nicely decorated, with the understated, elegant look. Our wine is similar. Another storm is on the way.

December 22 is the first day of ice wine pressing. After three weeks in a food freezer, the grapes are as frozen as can be. Inside each grape berry is a pearl of liquid not frozen; the rest is mostly frozen water. We look for weather just at the freezing point or a little above, so the grapes stay frozen all day. The slow pressure of the basket press forces the pearls of liquid out drop by drop.

I leave at 6:30AM to pick up the frozen grapes. Miguel helps me load the grapes. Later in the morning, one of the press bladders blows. That means we just lost half of the days pressing. We harvested one-third less crop, now we lose one press load, or ten gallons of ice wine, plus, a new bladder is $250 and takes two hours to install.

We have survived another year. Next year is our tenth year in business. They say that's where the rubber meets the road for small business.

2004

Friday, January 2 – I pick up frozen grapes again.

Saturday, January 3 – I'm up at 5:00 a.m. to pick up more frozen grapes I load the presses myself as Fred is too weak to help, and I man the presses all day.

I wrote a grant for solar installations for agriculture in Pennsylvania and we made the first cut for funding. The consultant and installer arrive to discuss options. The grant will provide $1.5M in matching grants for installations across the state.

Thursday, January 15: We bottle the 2002 Viognier. Cold stabilization is complete for all new wines.

Sunday, January 18: Snow is expected tonight. We have French parents of a customer come to visit the winery. The father grows table grapes in the cognac area and shares information about the vineyards in that region.

Monday, January 19: I work on the annual winery BATF (Bureau of Alcohol Tobacco and Firearms) reports. The Feds created a "sin tax" on alcohol after prohibition. They collect 20 cents per gallon on table wine and $1 per bottle of sparkling wine sold from each winery.

Fred says he thinks he should be in the hospital. He's been losing weight and has a lot of abdominal swelling. We spend the day getting admitted to Fox Chase.

Wednesday, January 21: Telephone calls come from the old guard wineries, who run the state wine association, and are so angry at me. The state's history is native grapes made into sweet wine, and they want it to stay that way. I got a grant from the Department of Agriculture to compete against them with dry wines and have broken up their monopoly. They are trying to strong arm us and are moving to form a competing Quality Assurance Program to put us out of business.

Thursday, January 22: Last night, Tashie tortured me all night. She didn't sleep, was hopping around the bed, whimpering at the critters outside. I could barely move her away; her strength is beyond her looks. There was no moving her off her watch. In the morning, she races out the door to check on the critter smells left behind. I check the back compost pile and it looks like there was a deer party.

Tuesday, January 27: Fred has been in and out of Fox Chase for the last two months. Today, we are told they can't continue more treatment. He can go to rehab to see if he can recover enough to take more treatment or go home to Hospice. He wants to go home.

Last week in January: The weather has been very confusing with five 65-degree days, and then overnight to 20 degrees.

Tuesday, February 3: Fred is worried about the vineyard spraying and is talking with me about that. We get out all the literature and discuss spraying issues, equipment, philosophies, chemicals, schedules, the idiosyncrasies of our vineyard, and how to handle the slope. He's worried that I won't be able to do it. I order the most current vineyard spray schedule from Cornell and start reading.

February 10 – Snow falls all day. Usually, we drive our old Caravan up and down the driveway to make tracks to get out to the road. Fred instructs me on how to use the snow blower. With a stone driveway you have to tilt the blower about an inch above the surface lest it pick up and throw the stones out. I last about an hour with the snow blower. I use a snow shovel several times throughout the day to keep the tracks clear enough to get out in case of an emergency.

February 20: There's been precious few hours to get pruning done. No time to worry about that now.

February 23: Fred wants to talk. He apologizes for telling everyone he was the winemaker. I say, "I knew about that, and it doesn't matter. What about the press rod flying an inch away from my head, the broken hornets' nest, and walking away from the ladder?" He hung his head and said, "I'm sorry, very sorry."

Thursday March 4: The solar grants have been verbally approved and we're ready to order the equipment.

Sunday March 7: Fred passes on, only 54 years old, after twenty months of a continuous struggle.

Thursday March 10: Jon and Barbara caravan with me to Valley Green, an old travelers' Inn on the Wissahickon Creek, dating back to colonial times, for a long-ago planned event.

Marnie Old, sommelier, leads the tasting. She is so magnetic and conjures up mouth-watering descriptions. Marnie comments that our wines are very French-like, both the champagne and the VSP (our Bordeaux blend), and tells participants, "You'll want to have some VSP in your cellar."

Friday, March 12: I discover an email about Fred's passing sent out by the state wineries organization. It reads as a crude announcement that he died and includes an invitation to his Memorial service, which is private. No one ever checked with me; inconsiderate beyond words.

Tashie sits up at night looking for Fred, waiting for him to come in from the vineyard.

Thursday, March 18: I spend some time with Miguel

in the vineyard checking to see if he's getting the work ideas. We walk the rows and check for various things; hidden shoots around the trunk, thin small shoots that were missed, canes too thin to produce, and old trunks to be removed. I tell him he is doing a very good job. "Thank you," he replies, with a beaming smile.

Saturday, March 20: It's a big workday with seven strong men, the John Deere and Kobota tractors, and three chain saws. Later, we have reclaimed ten feet along the upper tree line. One mishap occurred; two chain saws working on the same tree, and in their eagerness the tree falls on three rows of vines, breaking all the trellis wires and several vines. They all feel very bad, but I am not upset. Mom, Dad, and I have spent all morning fixing a lunch for twenty in the courtyard.

After lunch, I look for Tashie and am told she is walking up the vineyard hill. I go to find her; she collapses as I approach. I carry her to the house and prop her head up on a pillow on the couch. One week later, as I finish reading the book to her that I was reading for Fred during hospice, our kind vet comes to the house for her last breath.

The pruning continues and in late March there are reports all over the Mid-Atlantic about vine damage due to the January 65 degree highs. Some grape varieties were fooled by the warm weather and started pushing fluid up from the roots into the canes above ground. This is what normally happens in the spring when the vines start to grow. In January the five 65-degree days were followed by a 20-degree cold snap. In May, we learn that our oldest block of 500 Cabernet Sauvignon vines died. That was the only grape variety affected.

The new PLCB (Pennsylvania Liquor Control Board) Chairman, Jonathan Newman, was looking to recast the State Stores as more consumer-friendly and invited several wineries to participate in the new PLCB wine festival to be held in Philadelphia. Distributors from around the world would be attending. This was a first of its kind. The distributors were salivating for an invitation, since Pennsylvania was the single largest purchaser of wine in the world.

In early April, the first PLCB Wine Festival happened in Philadelphia, and ten Pennsylvania wineries are invited to attend, with over 200 National and International wineries. I struggle to put it together to attend but manage to get there with only some chilled wines for tasting. None of the other Pennsylvania winery attendees bother to visit my table.

Thursday, April 22: Five barn swallows flew around me as I was working in the Gewurztraminer. They chirped and swooped, and I said hello in a singsong attempt at bird talk. It's thirteen years since their first arrival, and always such a thrill.

Friday, April 24: I've been preparing for a new customer event, "Pruning Demonstration Workshop and Red Release Day." The 2001 red wines are our best ever and include our first Petite Verdot.

April 28: Our solar trackers are installed. We placed them in a garden setting. At sundown, they slowly rotate to face the east to be ready for the next days' sunrise.

May 10: I have completed three vineyard sprays with the new golf cart and thirty-gallon tank. That worked great with the small new shoots. I used a hand wand to direct the spray to the 15-inch shoots, instead of full tractor spray that would waste 50% of the chemical mix.

May 20: The new Cima sprayer arrives. Fred had been using a Hardi sprayer, which uses a "flood the vine" with spray material approach. It's a common approach; use 200 gallons of water and chemical mix per acre, spray the vines with the mix "flooding" the vine from the top, so the mixture runs down the vines, with 50% running onto the ground. The new mist sprayers force the spray mixture through small nozzles at 100 mph, creating a mist like spray, and it literally sticks to the vine leaves. These new sprayers are more expensive and require more horsepower, but you can spray twice the acreage per tank with no loss in chemicals.

After attaching the new Cima sprayer to the three-point

hitch and getting the initial instructions and overview, I calculate chemicals and water and adjust the nozzles to target the current vine growth.

Manzate, sulfur, and a foliar nutrient are the chemicals I use. Sulfur and the foliar nutrient are organic, manzate is a fungicide, but the older kind, not the new strobilurins. The strobs are the latest in chemicals, more expensive per tank but the spray lasts for 14 days. My routine is every ten days Also, the vineyard will become resistant to these new chemicals after a certain amount of material per acre. I'm committed to the most sustainable approach possible, with the least chemical input.

I filled up the sprayer for the first time with the maximum 80 gallons and went to the top of the vineyard. It was challenging turning the tractor, but the brush clearing we cut this winter gave me more room. On one turn, the front tires locked, and the tractor rolled forward toward the edge of the bank. I couldn't stop it, so I straightened the front tires, while the tractor with 600 pounds on the lift in back started down the bank at a steep angle toward the road. I hung on and was sure I was going to roll over, but the tractor remained upright. Maybe the calcium in the tires really made the difference. I was scared to death as I looked up to see three neighbors watching me, as I was all garbed up and looking like an astronaut. They waved as if everything was normal.

Saturday, May 29: Lucy just stopped over to bring me some irises. One was given to her by a neighbor friend, who died of cancer. She got choked up telling me. My emotions didn't surface quickly; I try to wait until our customers leave.

Monday, May 31: Today is Memorial Day, but Miguel wants to work. It's time to spray again. We are in reasonable shape getting the vine canopy up for spraying. I have Miguel practice driving the tractor with the sprayer on the back but not full, on the flat ground, and then on the hill. We get suited up in protective clothing and I fill the sprayer with chemicals and water. Then, the tractor won't go; it keeps cutting out. We check the fuel and Miguel thinks it smells like gas. We drain the tank and decide to work some new diesel through. We take the fuel filter apart, drain it, clean it, then rinse it, and fill it with fresh diesel. We rinse the main tank also, and finally fill it with fresh diesel. This works and the tractor starts and after a brief sputter keeps running.

Tuesday, June 1: There's grant money for agriculture; so, I draft an application for renewable energy for Pennsylvania wineries.

Friday, June 4: We work all day on trimming under the

vine row. It's a lot of work, but it eliminates the *chemical banding* (chemical banding in the vineyard is using an herbicide spray to create a 24-inch-wide area under the vine to prevent any vegetation from growing).

Saturday, June 5: Since it rains all day, I rack the Chardonnay, clean the floors, and label bottles in the winery. All the rooms in the house are cleaned preparing for the new puppy.

Sunday, June 6: I have a 9AM appointment to pick out the puppy, a choice among three females. When I enter the room one puppy jumps up, throws her paws on my shoulders and starts licking me. There wasn't much doubt about which puppy was going home with me.

Monday, June 7: Our chemical man hasn't returned my calls or faxes.

Wednesday, June 9–6:30 a.m. There is still no response from the chemical man. I call Brett to borrow some manzate. We put up shoots all day.

Friday, June 11: Top of the hill Chardonnay are already falling over the top wires. I cut back to 12 inches above the top wire and tuck the shoots up into the wires.

The Helena salesman still has not responded. I call the company's main office in New Jersey. They say he was on vacation, but is back now, and he should have let you know. I explain that my husband passed away and I am continuing the vineyard. I explained that I had faxed the chemical order to the salesman as we have always done but have gotten no reply. They apologize, but I say I want an apology from Ron, their salesman. A week goes by with no call, so I contract with another chemical company.

Saturday, June 12: I decide to take puppy for a golf cart ride up to the top. As I near the top of our hill, I see people on our property lifting stones from Fred's private stone wall. We dug these rocks out of the vineyard while we were putting in the hill Chardonnay. He used to periodically form them into a wall or cairn-like structure. I recognize the neighbors and their son. I call out "excuse me" and they ignore me as they walk over to their property and drop the rocks. I am shocked. Two of my neighbors and their son are taking our rocks!

I ask, "What are you doing?" They don't answer. "I can't believe you are taking our rocks!"

The man replies, "We are helping clear things out. Didn't you say you wanted to mow up here?"

"This is my property," I inform him. "You need to ask to come onto my property and you just can't take things," I shout.

"Well," says Olga, "I thought it was just a junk pile."

"Those rocks we dug out of the ground up here by hand

when we were planting the vineyard. This was a special pile..." I am practically crying.

I am livid and tell them to return everything and that there are other rocks missing. They deny everything. I leave, but as I get to the house, I am incredulous that they would take this liberty and I return to tell them I can't believe they would do such a thing.

"Is that what you would want me to do? Should I come on your property and look around and take anything I determine is junk?" I said.

The neighbors stand in place and look at me or look away and say nothing.

"I don't want to have to build a fence because I don't trust my neighbors," I call after them.

Sunday, June 13 – I decide to call the state police to file a complaint against my neighbors for trespassing and theft of property. They ask if I know who did it and I explain that I caught them red-handed.

The state police officer is a young woman; big, tough, and strong. She asks some questions. I just want them to stop coming on my property. She will give them a warning citation, but the next time will be a $300 violation each.

That night, we watch the night descend from the front porch, and the puppy jumps up high in the air at a pair of shooting stars in the sky. She will be known as Stars.

June 15: I decided that it was time to go and visit Lester Orrs, dairy farmer, and find out if I could ever buy his land. Since we came here in 1991, we have talked and asked him about the fields he owns next to our property. When you sit on our front porch, these fields look like part of our lands.

June 21: Summer Solstice Michael Clark drops by. He and Liz used to live down the street but now live in upstate New York. Michael was the president of the Natural Lands Trust and he and Liz are conservation consultants. He looks out at the overgrown grass everywhere and says, "Do you have enough help?" I couldn't imagine why he would ask this question. All that I see right now are the vines. You never have enough help with the vines. My grass strategy currently is to let it grow longer than normal, and it will slow down faster.

July 4th weekend: My dear long-time friends Mark and Anna arrive for the long weekend. Mark owns Lameroux Landing in NY. By now, the grass everywhere is about 16 inches and still growing. I can't get to everything. I just focus on the vineyard. Mark looks around and asks to use the Kubota and spends the rest of the daylight mowing everywhere. On Sunday, Mark starts mowing again and finishes late in the afternoon. We go for a scouting walk in the vineyard, and he identifies grape berry moth in the Chard, plus a little sunburn on the grape clusters, but they'll recover.

July 6: I notice more clusters with botrytis mold and cut off three trash cans full of clusters!

July 7: It's Wednesday. Dad has come to visit, after returning from Mount Washington.

The next day, I take Dad to see Lester to ask if he will sell me the fields next to our 13 acres. I explain that if I am to continue, I must find more land to grow grapes, since buying them is too expensive. I need more grapes to make enough wine to support the winery endeavor.

Over the summer I visit Mr. James, Lester's best friend, retired President of the Elverson National Bank to ask for help in talking with Lester. Mr. James calls back one day and starts with an apology. I'm confused at first until he finally says that Lester doesn't respond well to a woman about business. He tells me the story about Lester's mother and her overbearing nature. Lester had started his freshman year at Penn State, and one weekend his parents visited and there was a girl in Lester's room. Everything was proper as Mr. James was the roommate and was present. But Lester's mother was fuming and took him out of school. Lester never recovered and spent his life tending the small dairy farm.

Peg Welcomer lives across the street. Her husband died

during the years of Fred's illness. She has always been neighborly. I visit and ask for her help in talking with Lester. She agrees but says it will not be easy. A few days pass and she comes to visit and says that Lester will sell me the land.

July 10: I sprayed the upper vineyard, using a new strobilirium, Abound. These new chemicals are very expensive, but last for 14 days. However, they can only be used for two sprays per year; more may cause disease resistance.

July 12: I am preparing for another spray by tucking the shoots up so they don't get broken off by the tractor. It's difficult to keep up with the growth, and Miguel isn't helping by cutting the shoots too short. This will impact next year's cane selections.

July 13: I prepare for two tanks to spray the upper vineyard. The John Deere tractor radiator belt breaks. I saw a belt somewhere in the shed. I go up to see if Tom the mechanic is in, but he's out to lunch. I find the belt and it looks like the correct one. I get the manual and some tools and Miguel and I go to work. I'm the director and Miguel performs the repair. We both end up tugging and pulling and pushing to get the new belt tight. It works and we're back in business.

July 14: During the last tank of the last spray, Miguel

forgot to refuel and ran out of fuel on the hillside. I started to get mad and then caught myself, since something is always breaking. I got the manual; if you run out of fuel there may be air in the lines. I run back and forth getting tools and we bleed one screw—lots of air. The other screw seems stripped, and I can't find a tool to fit. We try bleeding the one screw again and the engine turns over!

These are the endless days of living on a farm.

July 16: There is something brewing in the hill Chard. It's the grey mold botrytis. I cut out all the affected grapes and trash them. I make a note about this recurring problem and the location in the vineyard.

July 17: I awaken at 5:30 a.m. and realize the scope of activity for today, a Saturday. Three vineyard workers report at 6:00 a.m., the portable sawmill at 8:00 a.m., the Westin auction group at 4:00 p.m., and Lynette to do the state store champagne tasting from 2:00 p.m.–4:00 p.m.

After ten minutes, the portable saw man has a scowl and spends 30 minutes telling me all the problems. I say, "That's why I wanted you to come look at the tree; I told you exactly how it is, now you have to deal with it!" In a nice way, I hoped. Finally, we got out the chain saw and started pairing the limbs off. The Stihl stopped automatically oiling the chain and I had to take it apart, clean it, and in the end oil the chain manually. I found a syringe and filled it with oil and that worked well every five to ten minutes. After several hours, the top section of the tree was trimmed and then we cut the 22-foot section in half. Ron, the portable saw man, told me he came because he was told it was a beautiful walnut tree and assumed it was straight. There's that trouble word "assumed" again.

I suggested using the slope of the ground and building a ramp up to the saw bed. I was sure the John Deere could drag the stump around, and it did after a few gyrations with the giant hemp rope.

As we were finishing the positioning, a customer came into the vineyard, having walked all around the house and grounds. She was holding a bag of half rotten white concord grapes that her husband wanted to know what was wrong. It looked like grape berry moth damage, but I certainly am not an expert on native grapes and referred her to the county agent. I was too busy to be mad about the thoughtlessness of bringing diseased grapes to my vineyard, until later.

August—The night patrols start, and new puppy Stars loves this job. I shine a flashlight down each row looking for deer. If we find one, we chase it out of the vineyard. First patrol is at dusk, second before bed, and third sometime in the middle of the night to keep them off guard. We used to put radios out, dialed into talk stations, and move them around the vineyard every two days, covered for rain, on from dusk to dawn. The theory is that the deer thought people where present and stayed away. This worked for years, but this year I found a doe and two fauns eating the grapes right next to the radio one morning listening to Sports Talk!

The other job for August is emptying the winery tanks, barrels, and bottling to make room for the new harvest. This year we find ourselves with little red wine in the cellar, because of last year's poor harvest and the winter damage that is translating to low yields this year.

Before Labor Day, the state viticulturist sponsors a seminar in Adamstown at the Penn State Ag Research Center. The agenda includes calibrating your sprayer, soil nematodes, and touring the experimental vineyard plot. Ron, the Helena Chemical salesman who never returned my calls is present. I glare at him. When questions arise about spray timing, he pushes an every 7- day schedule, except for the new generation of expensive chemicals that are needed only every 14 days. I say that we have been using the recommended schedule for a 10-day cycle, for years, with no problems. There is push back among some attendees. Later Dick Naylor, from Naylor Wine Cellars in southern York County, tells me he uses the same 10-day schedule, and that he's been trying to tell them at Penn State for years, but they won't listen.

In late September, Lester and I agree to the terms of sale, dependent on a soil sample. I have been learning about the Land Preservation Grants available through Chester County and meet with the Township Administrator to make application. With the new acres in vineyard, the two plots will qualify for the selling of the Development rights and pay for the increased mortgage. The Township Administrator says that no one in northern Chester County has received this grant, and that they would whole heartedly support my application. Then they added "and you're actually farming."

Chester County Day, October 2–The Welcomer's across the street are always on Chester County Day which hosts the oldest "old house tour" in the country, as well as a benefit for the Chester County Hospital. Five thousand tickets are sold for this event. This year, I offered to be a tour stop for them, with a port-o-potty and water available. I fixed the walls in the 1700s log cabin and opened that part of the house for the tour. We had many new wine customers.

During the fall, I focus on finishing the window painting on the winery addition. Fred had opted for the weathered look, and now I wanted to create a finished look. One of the Pella windows installed in 2000, and visible from the road, had started rotting. Karl had the Pella rep come out last year, and nothing happened. I call the local office and get the new salesman to come look at the problem window. He will get back to me about what they will do.

November 7: The Quality Alliance wineries are coming today. Twenty wineries will bring their dry (less than 0.3 % residual sugar) European Vinifera wines for a group tasting and evaluation. There is a lot of conversation about the different wines and personal preferences. The judging panel will approve wines, and those wines will receive the first Pennsylvania Quality Alliance stickers, like the well-known Canadian VQA program.

This year, I have written and secured a small grant for the group to showcase these wines in selected Pennsylvania State Stores in December. But again, the State only purchases one case from each winery, and we must provide and stock a display. I manage to get a representative from each winery to set up the display and their wine for sale in the State Store near them. I encourage the wineries to do an in-store tasting. Many complain that they thought I would do that for them.

Some of these wineries are from Chester County and talk about their new Chester County Wine Trail. I inquire about joining and am told that the trail map with events for 2005 has already been printed. I am told to try next year.

After Chester County Day, I am asked to be on the township's Christmas House Tour. My wine tasting couples agree to help that day, so I can open the house and the winery. We have many local visitors who have never been to the winery since we opened.

2005

New Year's Day—The last two years have been very difficult due to Fred's illness. I can't imagine what to do next, or what I can do. "We" is now "I."

In late January, I call the Pella regional office and ask for the manager. I'm mad and want action. The rotting Pella window faces the customer parking area and is very unsightly. The manager comes out to see the problem window with an assistant. After a half hour of listening to them talk about all the potential reasons for this window rotting, and none are their problems, I throw them out. That night, I search my records for the original Pella receipt, since I keep everything. The warranty is for 7 years against defect. I call the manager the next day and read to him the warranty. The replacement window is delivered the following week.

In February 2005 the purchase is complete for an additional 19 acres, about eight of which are tillable. The fields are covered in brush since they haven't been mowed for years. The old John Deere tractor proves its worth again and we slowly knock down the growth. The brush mower is five feet wide, but I work on three feet of width at a time, because the brush is so dense. I marvel about the thrill of working the tractor. It's hard and challenging and takes concentration, but very rewarding as the work slowly gets done.

Dave, my sister's husband, has been helping me this year. It's been a blessing; he's reliable, trustworthy, and very skilled. Since Fred was in treatment, I couldn't finish any champagnes. Now, I have taught Dave, and we are making good progress building up a little inventory for sale in the fall. When we start the 2000 vintage, we experience bottle breaking at the neck in the corking process. It's start and stop without a clue as to the problem. After two days, we have 10% breakage. The bottle distributor and manufacturer say they have no reported problems. We contact the cork supplier and the corker company with no result. We try to continue slowly lest we are inadvertently causing a problem. After two weeks, we decide to stop and examine all bottles in the vintage. We discover that each bottle is numbered for the batch and mold on the bottom rim; all the broken bottles have the same number on them. We relay this information to the supplier. After some haggling, the manufacturer's rep agrees to make a visit. He asks to see a demonstration; the bottle with that mold number breaks immediately. He tries to say there could be many reasons. I persist, and he finally admits that it is a bottle mold irregularity. I say we want compensation. He acts surprised, and after some pushing, agrees.

The first plantings in the new vineyard will be Petite Verdot, Syrah, and Cabernet Franc. Amberg's nursery in upstate New York will try to supply our order. They know about Fred and have been very kind and complimentary of my fortitude in pressing on with the vineyard. We end up planting four acres, and then struggle with the excessive weed and bramble growth.

Late March–PWA Annual Meeting, Harrisburg. Most of the 120 Pennsylvania wineries and some from neighboring states attend the annual meeting. There are speakers from the Department of Ag, the Pennsylvania Liquor Control Board, The State Tourism Bureau, Cornell Viticulture, and other invited guests. This year, I have been asked to talk about the new Wine Quality Alliance.

Also, I recently received word that my grant application for renewable energy for Ag in Pennsylvania is approved and funded. It will provide $1M in grants to the wineries.

I explain that the Pennsylvania Wine Quality Alliance was designed to gain attention for the dry European

Vinifera wines produced in Pennsylvania. Most consumers think first of California for these types of wines and were unaware that Pennsylvania also produced many of these wines. In other states (e.g., California, Washington, New York, and Virginia), these types of wine increased the region's overall reputation for wine; increasing winery sales, and local retail sales. A 2001 study, for example, indicates that every $1.00 of Washington State wine sold at wineries, retail stores, and restaurants generated $2.05 of total economic impact for the State. I explained that most Pennsylvania wineries produced wines from Vinifera grapes and that all wineries could benefit from an increased interest in Pennsylvania wines. At that time, the majority of the million gallons of wine produced in Pennsylvania were not dry Vinifera.

During my presentation, half of the attendees began to shout me down and yell that they will never allow this in Pennsylvania! There were a few wives attending, but none spoke up. The men kept it up so that I couldn't finish my presentation. I was mortified and had to fight hard to resist running out of the room.

April—A letter arrives from the County regarding my application to sell the development rights. They say my application is still incomplete and will be set aside for consideration in the following year, due to the fact that my land has not been listed in the township's Agricultural Security Area. When I follow up with the township, they say "Yes, you are in the Ag Security Area."

The replacement Cabernet Sauvignon vines, from the 2004 winter kill event, will arrive at month's end. I devise a way to remove the remaining vines using the John Deere tractor and big chains. The trunks are four inches thick. I set up to pull the vine out and discover the root systems are 15 feet deep!

May–This year, the PLCB is expanding its Wine Festivals to Harrisburg and Pittsburgh in addition to Philadelphia. They have purchased ten cases of our Blanc de Blancs and will have them at the festival store for purchase. I will have to go alone to Philadelphia but will have help at the other two locations. I have a small folding cart used for the State Store Tastings. I decide to take the framed Vinalies award, cooler with chilled champagne and ice wine, a vase and flowers for the table decorations. This year only six Pennsylvania wineries were invited. My table is near an entrance and across from Gallo. The Gallo booth is enormous and has constant traffic. Gina Gallo is the festival's most famous person and has distributors, salesmen and attendees hoping to meet her. Their booth shows glorious pictures of Gallo vineyards in California and Italy. Fresh long stemmed red roses are delivered several times a day to keep a large basket always fresh looking. I am hoping to meet Gina and notice that she is looking at the Vinalies award. When there's a lull in the traffic, I walk over to her booth, and she suddenly disappears behind a screen. On Saturday, Jonathan Newman, the PLCB Chairman, stops by the booth for a photo op and there is a little interest in my wines. Later in the afternoon, Gina is alone at their booth, and I approach to introduce myself. Again, she doesn't look up and disappears behind their screen. I don't try to meet her again.

That December, there is a big spread in the industry news that Gina Gallo is heading up a new brand for Gallo for premium wines. Two years later, Gallo receives its first Vinalies award, and there is much ado about it everywhere.

Barbara and Jon drive across state with me to the Pittsburgh PLCB festival. Pittsburgh was known as the "Gateway to the Midwest" and full of friendly, happy people, the descendants of the great steel town. Of the three cities in Pennsylvania for the wine festivals, this one has the most consumer interest at our table.

On the way home, Barbara talks about the Boston College project and how they want to build a new system to help Medicaid recipients select and manage their own care. The big consulting companies have all proposed multi-year enterprise systems with huge costs. I ask my usual questions to discover what the critical success factors are. I share some wisdom from 25 years in the I/T (Information Technology) business, that a successful I/T implementation (putting in a new I/T system) should be done in one year. Many things can change after one year. It's best to carve out what can be accomplished in that time. Barbara asks if I would talk with her boss about that. In July, I agree to perform a small needs assessment engagement that would identify a possible scope of work that could be used to evaluate proposals.

Summer–One night in August, while I am sitting alone outside on the front porch enjoying the breeze and the sky view, a car pulls up to the driveway barrier, which is a simple sawhorse serving as a "we're closed" message. The house is dark. I often work in the garden or just sit outside till dark, and then go inside when it's time to go to bed. I watch as someone gets out and walks around to the front of the car. The car is still running with the lights on. I think maybe he needs to use the port-a-potty, which is near the road. He peers at the sign with our hours and stands for a moment looking around. Then he returns to his car and gets in, but not to drive away. He turns his car off and the lights off. My hair bristles and I look for Stars, who is not in view. She's running in the vineyard. The front door of my house is propped open. I quietly move inside, grab the .22 rifle and some shot. I slip through the open door and stand in the front walk watching. The man walks slowly down the drive, pausing to look at the winery doors. I am in the dark unable to be seen. He walks up to the winery door and tries to open it, but it's locked. He continues to the next section of the barn and tries that door and comes to the door closest to the house and tries that too. My night vision is good, and I follow his outline as he starts to move closer to the house. That's it, no closer. I point the gun at him and say in a loud, firm voice "We're closed." Startled, he stops. After a few moments he says, "Oh, I just wanted to pick up a bottle of wine." Right. "We're closed!" I say again in a firm voice and move down the walk closer, still holding up the gun. "I've been here before, and my wife wanted me to pick up a bottle." I move to the end of the house walk. "Our hours are listed on the sign you looked at by the road. You'll have to come back when we're open," I say, loud, firm, and final. He hesitates and then says "Ok, sorry" and turns to walk back to his car. I watch as he walks back to the car, gets in, starts the car, and backs out of the driveway. I watch to see that he drives back up the hill and continues out of sight.

I sit in the dark on the porch with the .22 for an hour. It feels safer outside than inside. I must get a gate.

The birds are a big problem at harvest, and especially just before harvest. They peck at the grapes starting around the end of July to test the sugar levels in the grapes. They are a big nuisance as they fly down to peck one berry on a cluster. That berry then starts to rot, and eventually affects the whole cluster. Fred would patrol the vineyard with a shotgun, firing into the air several times per day to scare them away. I started the effort to get the entire vineyard under bird netting years ago, but before the nets are on, the birds can still "test" the grapes. There are sound machines that produce a loud boom sound that is very effective, but we live here and are concerned about our neighbors. I decide to try bird scare balloons and a sound machine that has various bird distress calls. This sound machine is very effective in keeping the birds away. I use it in the first vineyard area to be netted, and then move it to the next area, as each variety is harvested. The balloons look very festive, the visitors comment favorably, but it is not effective to scare the birds away.

Harvest comes and goes easily as the weather has been ideal, not much disease pressure and warm dry days for maximum ripening. This year, I have selected new French Oak barrels for the Burgundian Chardonnay. I am confident in my barrel fermentation techniques, preferred yeast, and the very important malolactic starter additions (malolactic starter converts the malic acid [more tart taste] to the lactic acid [more soft and buttery taste]). These are important steps to making a fine Burgundian style Chardonnay but adds costs and time.

I have enjoyed my work with Boston College and met their national team of the states participating in this endeavor called the Consumer Directed Module. I helped them identify a Project Scope, evaluate vendors, and establish a Project Management structure for the project. They have asked me to continue as the outside Project Manager. They offered me a full-time position to work remotely. I didn't think it was fair to accept the offer when I knew my attentions would be so divided, trying to keep the vineyard and winery going. The project will start after the New Year, and I will work part time.

For this year's Champagne Day event I decide to include Lancaster County cheeses for the guests, with breads from St. Peter's bakery, as well as fruits and dips. It's a beautiful day with attendees enjoying wine tastings and music in the grass.

Two weeks later I get a letter from the Chester County Department of Health saying I will be fined for serving food to the public without a license. I call them and they say it was reported that someone dropped a plate of food on the grass, picked up the food and served it. "This was a private event—by invitation only," I say. They reply that it was reported as a public event. "Who reported it?" I ask. They say that information is private. I say again, "This event was a private customer event—so it must be a customer who reported it. All attendees must present their personal mailed invitation card to be admitted." After some stammering the person replies "Well, one of our employees was in attendance." I reply, "Then you already know we were not serving food to the public." Later that day I search the Chester County Department of Health employees list and delete the culprit from our customer list.

The red grape harvest is very late. It's important to have full ripening time as that affects the finished wine flavors. The early ripening flavors show in the front palate, mid-season ripening flavors show in the mid palate, and the late season ripening flavors show in the back palate. Sometimes, you will hear a wine described as having a "long finish," which means in the back of the palate, and down the throat. So, to have a "full" red wine you need full ripening. We end up fermenting the reds into December. The ice wine grapes are in the freezer, and we wait for colder weather to press.

I send a letter to the Chaddsford Winery owners who are running the Chester County Wine trail asking to join and follow up with a phone call. I am told that the members will be taking a vote on my membership request. I call each of the winery members' owners: Stargazers, Twin Brook, Va La, Kreutz Creek and Paradocx to ask for their vote. "Why wouldn't they include us? We are in Chester County," I say to each owner.

2006

We start after the New Year and travel every day to Denver Cold Storage in Lancaster to pick up the frozen grapes. After three weeks, we are finally done pressing and the juice is safely fermenting inside.

In late January, I call Tony at Va La to ask about the wine trail. After much pressing, he says they won't have a vote on our membership because Chaddsford won't allow it.

The winter proceeds without major snow. My 4-hour battery pruning shears starts having problems. I switch to use Fred's 8-hour battery pack, which is significantly heavier. Since this is my hardest job right now, I always use the morning hours when I have the most energy. The pack weighs about 25 pounds, but worth its weight in gold for the work it can do.

In late April, I check each of the old Cabernet Sauvignon vines that survived the 2004 winter kill—about 20%. Some vines have new buds pushing out of the graft union. I leave the old dead trunk as support for any shoots that emerge. This area I refer to as a renewal zone. If a vineyard is banded with herbicide, this area is kept "clean" so new shoots can't grow. It's a lot of extra work to allow new shoots to grow from the base of a trunk, but there are many benefits. The old vines have established root systems that bring fruit to ripen faster than young roots. The old vine doesn't have to be removed, and a new vine purchased and planted.

The new vines were ordered last year for the second half of the new land. This new half is a very steep hill. I carefully lay out the new rows from the top: Chardonnay, Merlot, and Malbec. Malbec is the fifth of the Bordeaux grapes, and this is the first time I will experiment with this variety.

I follow up on the application to sell the Development Rights, and a county clerk says that the Township never recorded my land parcels in the Ag Security Area. This time the Township administrator says that they just filed that recording with the County. I'll have to keep paying the second mortgage.

Easter Sunday Mom and I prune and clear the Merlot and Cab Franc vines. A carpet of violets greets us. Glorious, beautiful, nature saluting us. We take pictures and feel like royalty.

In 2002, I had converted the vineyard to permanent under-the-vine row cover. This eliminated the herbicide banding. We have a lot of red and white clover which can out compete the unsightly weeds. It wasn't perfect, but another step in eliminating chemicals.

May: This new Kubota was one of the first compact utility tractors that featured an under-the-belly mower, a bucket, and a backhoe. This would replace the push mower I used to mow around the house, and could mow the vine rows much faster than the John Deere. The Kubota produced a "finished" mower cut with two passes in each vine row.

The new vineyards doubled the mowing job, but I was determined to keep things looking kept up, so I mowed till dark every day. One evening, I was heading back to the house, and wanted to finish mowing around the honeybee tree. I had yet to remove the pallet behind the beehive that was a winter windbreak, and the Kubota bucket caught the edge of the pallet and toppled the hive over. The three "stories" (called supers) housed about 50,000 honeybees. The supers contained honeycomb filled with honey and weighed about 50 lbs. each. The bottom super was attached to the "bottom board" of the hive (like a house foundation) with propolis, a substance the bees make to seal cracks. Slowly, the super comes loose with a lot of prying, from the propolis, and I set the hive supers back in place. NASA has been trying to make a synthetic propolis or get enough honeybees to make enough to seal the space shuttle!

Planting the new vine rows was going very slowly in this area full of rocks. As we worked to plant a row, we gathered up buckets of small stones, 3-inch to 6-inch size and transported them to the detention basin area. I asked Miguel to bring two other workers with him to help with the work. Spanish was their main language. I know only a little Spanish, and the technical terms for work in the vineyard were difficult to explain, so I start to write instructions and translate them online. I read the instructions out loud in Spanish as they follow along with printed instructions. I follow up with work demonstrations and then ask each to perform the same. Time goes by quickly with little accomplished. The next day, I decide to break the work up into smaller work items and divide the day into three sessions: first morning, after morning break, after lunch. This works better.

Another letter from the County says that my application to sell the development rights is still incomplete and advises to make a new application for the next year. This time I write a letter to the township requesting a written response that my parcels have been filed in the Agriculture Security Area.

June: Stars was exploring on her own too much. I retrieve her from the farmer's dairy barn down the road one day, and decide I'll have to tether her to me to get work done. That evening, I ready the tomato plant and herb area next to the house. She tolerates watching me plant the tomatoes, but behind my back she shreds the basil plants. The next day after our walk, I attach her to the vine row near my work area. She is quiet for a long time, and I glance up to make sure she's ok; to find her standing with about 100 feet of line wrapped around her body and face and staring at me. "Oh!" I gasp, as she jumps in the air and the line drops free to the ground. How did she learn this trick?

I decide to give Miguel a space for a tomato garden this year. I got him 18 plants. Lynette asks "why?" I reply, "So I can have some tomatoes this year!" All of the workers and visitors always look at my tomatoes longingly and say "Oh, tomatoes!" so I give everyone a bag until they run out. Now the workers can have their own—it's self-preservation! On Thursday, Miguel stays after work to plant his tomato garden. I'll keep an eye on it over the weekends and water.

Before I left IBM, I used to ride my bike about 100 miles per week, 20 miles up and down the hills in our valley and a long ride on the weekends with Patty and Joe, my very special friends. This year Lance Armstrong is hosting rides in various U.S. cities for the Live Strong organization. I've done many such rides in the past and asked Joe and Patty if they would ride with me. The ride has a special entry jersey for survivors and caregivers.

Monday, June 5: Stars and I are sleeping past sun-up this year. She gives me a little tap on my feet and then starts nibbling my toes.

I put my socks on before getting out of bed; the only way to get them on before noon. In the winter, I just sleep with them on. The RS Medical Neurostimulator unit is next. Four sets of two pads positioned on my lower back. Press the on button and set the intensity... ahhhhh, I instantly feel better.

Every day, I am aware when I walk into the kitchen that I am walking back into time, where there are the memories, laughter, tears and hard toil of the settlers before me. I put the kettle on, get the Melita coffee filter out and check for fresh coffee grounds. On the way, I let Stars out the front door and prop it open with an empty plastic peanut butter jar. I was baiting the ground hogs with peanut butter in the traps.

Then I take one half of a 4-mg. cyclobenzaprine, a muscle relaxer, with a swig of raw milk. There is practically no disc material left in six cervical vertebra and four lower lumbar vertebrae from an accident in 1980. My neck and back muscles are in a constant war, tight as a drum. A major spasm is always a breath away. These things help to keep me up and moving every day.

July–The Japanese beetles are a growing pest this year. They usually arrive for a few weeks and eat the grape leaves from the top and move on. By late June, they have eaten nearly half the grape leaves. I resist quick reaction to this problem, since it will mean insecticide spraying that will harm all insects–including butterflies, honeybees, and hummingbirds. Finally, I have no choice. I decide to spray at night when the butterflies, honeybees, and hummingbirds will be in their nests. The lights on the John Deere work and guide me through the vine rows. I spray only the fruit bearing vines and complete the task in two nights. The young vines will recover enough leaves from the Japanese beetle damage to prepare for winter.

We continue the hand work in the new vineyard this week, removing stones and small rocks, and trimming close to the new vines. I'm committed to not using herbicide banding in the new vineyards.

Last week Juventino, Tulio, and Gerardo were trimming the young vines and Miguel was spraying during the morning. At mid-morning, I go out to check them and Tulio is not with them. They are taking a break and

Tulio is nowhere to be found. I look for him and he has seemingly disappeared. I am aware that in general the workers have been taking very long breaks and always out of sight. So, I check them frequently. I have been told that one worker repeatedly sits down when I leave and rests the whole time.

Wednesday, June 7: We repeat the work plan in the vineyard, but progress is slow.

The Boston College project has my attention. We work remotely, using teleconferences, email, and web casts. Three states will install the new software. This project is a blessing; they appreciate me, and I appreciate them.

Friday, June 9: Lynette heads out to a farm visit for the Energy Farmers Project (the solar grant for Agriculture project I received from the Pennsylvania Office of Economic Development). Each farm is evaluated for solar installation capability, and a proposal for that farm is developed. The grant will cover 50% of the costs.

At mid-morning, I check on the workers and their work. Miguel's eye is slightly red and Juventino (Miguel's brother) asks if I have something for his eye; he thinks he may have gotten poison ivy in his eye. I look at the eye and ask everyone to stop. I explain about poison ivy, pointing it out in the groundcover and its three-leaf formation. I repeat frequently about the poison ivy; don't touch it, wear gloves and long sleeves, wash your hands and face after working, before lunch, and again at the end of the day. I go back to the house and make a saline solution, boil water with salt, get some cotton balls, put everything in containers for Miguel. Before I give the workers their weekly pay, I review the precautions and take them to the locations outside and inside where they can wash up. Also, I say that I think they have been taking too long for breaks; they should have gotten the trimming finished. As I hand them their pay, they are hanging their heads.

Stars has an abrasion between two of her toes. She was watching the giant backhoe go up and down moving dirt and rocks yesterday in the new vineyard and would try to dig in the rough dirt. I clipped the fur and cleaned the area, but her paw is swollen now. After the workers leave, I call the vet and get the answering machine.

The roofer, carpenter, and stone mason arrive at 5:00 p.m. to discuss the main barn roof replacement. The 1830 stone barn has the remains of a tin roof dating to the early 1900s. I'm concerned with how it will look from the inside with the one by twelve-inch boards, the new verge boards, the hump in the roof line, and pitch transition. After agreeing to the schedule, we discuss the stucco removal on the remaining stone wall that faces the house. Rolland will bring his scaffolding and compressor and air chisels.

He will teach the workers, and they will get a significant hourly rate increase for this specialized work. Instead of paying a contractor for his laborers, I try to offer my workers the chance to learn new skills and receive higher pay. Everyone seems to be happy.

The vet's call service leaves a message to go to the emergency vet in Trooper, Pennsylvania. Stars and I leave right away. She has a puncture wound. We're home by 11:00 p.m. with pain meds and antibiotics.

The weekend is spent "stuffing" shoots up through the trellis wires to prepare for spraying on Monday.

Monday, June 12: The vineyard workers arrive before 7:00 a.m. Miguel will work with me on the vineyard spraying this morning, while the others tie up the replacement Cabernet Sauvignon young vines. These are the new vines replacing the 2004 winter kill. I'm mad from the long breaks the last weeks, and it comes through. I try to change my voice tone as I direct them to finish what they didn't finish on Friday. The roofers arrive late morning and ask if we can unload the wood when it arrives. I say I can't unload it alone, so it must be before the workers leave at 4PM. After lunch we cover up everything inside the barn, clear the outside courtyard, and secure the equipment. The wood arrives at 4:00 p.m. just when the workers were leaving. They stay late to help.

Tuesday, June 13: The roofers arrive at 6:30 a.m.. First, they tear off the tin, trying to pull the sheets off to be salvaged. Then, the old cedar shakes crumble down, creating clouds of dust everywhere. Then, the furring boards and lathe-like strips are pulled off, with old nails sticking out everywhere. All afternoon we haul debris, filling trash cans for dumping on our burn piles.

The barn is wide open to the sky. The old rafters hang in the air, by their own force they hold up the roof.

More mowing to do tonight; I must keep up.

Tomorrow, Mark Chen, the Pennsylvania State Viticulturist, is stopping by for a "surprise" visit.

Wednesday, June 14: The one-by-twelve boards are going up quickly. I climb into the top of the barn, and don't like what I see. Daylight shows through with big slits between the boards, some slits as big as an inch wide. There's a two-inch gap where the roof transitions.

I call to Tom to stop, but his workers keep going. No one bothered to look from the inside; even though this was discussed at our first meeting. I insist that work stop until Karl arrives. Tom tries to convince me it won't be noticeable when the black roofing paper is on, and you can't see daylight. Then he says, "Well maybe you can get someone else to do the carpentry." This is a standard contractor approach. I know they're not happy, but neither am I.

Karl arrives at lunch time and agrees with me. They should have done a check after a few boards were up. I agree to accept what's up there, but the rest of the boards go back to the mill for a straight cut.

The roofers have gone on to another job. They promise to be back on Friday.

We stay late to load the wood onto the mill truck. I give the driver a bottle of wine.

Mark Chen arrives at 6:00 p.m. and is impressed with the new vineyards. We chat while walking through the different vineyard areas. He comments on the Vidal being weak looking for a hybrid grape variety. There's some discussion about the "own-rooted" hybrid grape varieties becoming susceptible to nematodes in the soil. It could be from tomato ring spot virus, which can be tested for. I had collected wood from the Vidal this winter to have grafted by Herman Amberg for planting next year in the new vineyard. I'll have to test that wood before grafting.

Thursday June 15: Tom calls at 6:30 a.m. to say the wood is being returned this morning and can we unload it.

Rolland arrives to give instruction about taking off the stucco. I attend with the workers and check that everything is safe. We nail plywood onto the roof and a separate foot stop.

Each person gets a training turn with the stucco removal chisel, and all seem very interested. The day before I have read a translation to them about the new work, and that they will receive 35% more per hour for this work. Three of them are anxious to start. I organize them in teams, and stress safety. Miguel and Juventino start first. They want to keep working and stay late. I limit them to four hours per day, including the collection of debris to fill in holes on the lower field road.

Friday June 16: The roofers are back and finish the front of the barn with the newly milled wood. The result is better, but I still must check every five feet to keep them focused on closing up the spaces between boards. After lunch, we help move the equipment to the back of the barn, then start the cleanup in the front courtyard.

I am on the phone off and on for the BC project, and in between, I check on the roof project and keep the vineyard crew moving along.

Tuesday, June 20: That's it. I've had it with Mario, the roofer's nephew. Last Friday, Mario came to ask where to dump the roof debris. I showed him where to put it, so it will be a "safe" burn. Later, as I am checking on the roof work and vineyard work, I find that Mario didn't put the debris where I had shown him. We will have to pick up everything by hand and move it. I tell Tom saying, "That's not acceptable." He apologizes and says he didn't know. I say I told Mario.

Wednesday June 21: In the evening, I take my bike out to Amish land to this wonderful bike shop, Shirk's, for service. Everyone is dressed in grey and black, and the barefoot wife and daughters are fixing tires and building new bikes. I need a new rear chain and new tires. While waiting, I pick out wild handlebar tape to match my Bianchi green bike, which was purchased in Port Chester, New York in 1988. Luke, the owner suggests a younger man might be able to advise on the handlebar tape suggestion. A big, handsome young man looks at the Bianchi and then at me and says, "That's your bike?" Yes, I say. The four other bike technicians offer their assessment of the handlebar tape choice: "that's what we call so ugly, it might actually go with a Bianchi!"

Thursday, June 22: The roofers finish the new roof. It looks like a new barn! We'll start the inside cleanup tomorrow after we chase that darn starling out.

Friday, June 23: The vines have grown to 80% of their size in the last three weeks. We have managed to keep up with the growth. All hands working in the vineyard every minute, every day.

Saturday, June 24: With rain over night, the leaves are too wet to spray. Miguel is here at 6:00 a.m. We walk up and down the Vidal and Gewurztraminer rows picking up shoots and tying them out of the way of the tractor. It rains at 9:00 a.m., so we cancel the spray again.

Sunday–More rain falls all day.

Monday, June 26: It rains early, but clearing is the forecast. We have to spray since the grapes are "naked" (without spray coverage). We start spraying at 10:00 a.m. It downpours at 11:30 a.m., and we abandon spraying for today. There is a powerful stalled front with moisture feeding from a tropical storm belting the northeast. It rains all night.

Tuesday June 27: Again, it rains all day. I decide to spray tomorrow rain or shine. The theory is it's better to get the spray on the leaves even if it rains right away. I empty the sprayer tank again into 5-gallon containers and rinse the sprayer and nozzles.

Wednesday, June 28: It rains early, but we start spraying in light rain. The tractor breaks down in the field. One of the sprayer belts is shredded. It's not something I can fix. The equipment dealer in Ephrata says they have to bring it to the shop, and agrees to pick it up this afternoon, and work on it tonight. They heard me say "my crop is at risk." They should have it back by noon tomorrow. God bless the Amish shop.

Thursday, June 29: The tractor is back by 9:00 a.m. We

start spraying immediately and take shifts all day. It's 2½ hours per spray tank, then clean and fill, and start again. I finish around 6:30 that night.

July 15: When Fred got sick in 2002, we had to stop all activities that were not crucial, including all maintenance. We had a list, but it just had to wait. In January 2005, the year after Fred's passing, six out of seven roof sections between the house and the barn/winery leaked. We just completed a new roof on the main barn and have five sections left to replace. All manner of equipment and appliances need repair. Some are very dramatic in how they bring your attention to them, like the kitchen glass cooktop that exploded one day while cooking on it. The glass shattered like a car window and exploded out horizontally. The manufacturer's rep wouldn't believe it and insisted that I had dropped something on it. It took six months before I could locate a replacement that fit the space. One piece at a time the glass fell away, and I was left with only one burner operating when the new one finally arrived.

Over the last three days:

- One van has no AC or gas gauge, and the electric locks don't work.
- Both vacuums are broken.
- The oven takes two hours to reach 375 degrees.

- The newly fixed vineyard sprayer has a slow leak and must be watched.
- The Kubota blows out black smoke after changing the engine oil, and the hydraulic fluid leaks somewhere.
- The front porch deck is about to cave in.
- I had to fire the accountant who kept saying she'd get the taxes done in a few weeks, and it's two months later.
- The propane ran out last night as I tried to cook dinner at 9:00 p.m.. Why? The company decided we didn't use very much so they would only deliver once a year.

The invasion of Japanese beetles in the vineyard is unbearable. I've been mowing the last two nights to get ready to spray. I have beetles everywhere in my clothes from putting up the shoots. Stars tries to help and eats them as I shake out my clothes in the front yard.

July 1: On Sunday morning's scouting walk, everything looked "clean" (no disease evident). Sunday night when I was mowing, I saw light brown berries in the Chard. How could that be? Saturday night it had rained, but disease would take several days at least after a rain to appear.

Last Monday, we sprayed with a very good chemical against botrytis added into the regular chemicals. Botrytis is what I thought I would have problems with. The grape berry moth burrows into the developing berry, then eats the grape from the inside, pops out, and dives into another berry. This starts the rot which usually turns into botrytis. I pick out the affected berries each year, but this is much more than usual.

I check the Cornell publications. It may be black rot, which has a 2 to 3-week incubation period. That would coincide with the tropical storm.

Monday July 20: The annual bottle delivery is fraught with problems; no loading dock or bottle storage area. We're space constrained and always have to clear space to take in the pallets of bottles. Two workers plus me are needed to prep the space for the pallets, and to manage moving the pallets inside the winery building. At 6:30 a.m. Miguel arrives, but Juventino was a no show. I was hopping mad and yelled at Miguel. "That's it!" I bellowed. "I've had it with him. He's fired!"

We have two truck deliveries: ten pallets of bottles in one delivery and the other delivery is six steel cages for champagne bottle storage. We have to be ready and get them unloaded from the street, as quickly as possible. We start to work on the preparations. Miguel works extra fast and pays close attention. Lynette arrives and helps to finish the prep. We're ready by 8AM. Steve arrives a little before 9AM with the forklift on his big tractor.

One truck driver calls—he is leaving New Jersey and should arrive before noon. At 9:15, the steel champagne cages arrive from France. At 11AM, we have not heard from the other truck driver, and call the trucking company. An hour goes by, and no one calls. We call again and they say they haven't heard from the driver yet, and he's not answering his cell. At 1:30, we call again. They can't reach the driver and say here's his cell number; you can call. There's no answer. At 2:30, I call the trucking company. They just heard from the driver and he's in Morgantown, just off the turnpike. "Why did it take so long from New Jersey? It should be two to two-and-a-half hours?" No one knows. Morgantown is a ten-minute drive. An hour later the truck arrives.

Finally, we begin to unload. They said they went through Philadelphia, where they must have stopped for over three hours when no one could reach them. It's almost 100 degrees and the road starts to melt as we unload. There's a trail of tar the length of the driveway, from the street where the truck is parked to the first door of the winery. We do a good job and get the truck unloaded and on its way before long, and then work at getting the pallets into the winery in the right order. We make quick work of it and look like pros.

Before Miguel leaves, I give him a letter for Juventino

that says he's suspended for two weeks. I feel badly for Miguel because I know it will put a drain on him.

July 20: Ike called around 5:30 last night. "It's hail damage," he says. That's the only thing that would split the berries. He describes two kinds of hail; one that's round and smooth and sometimes has liquid on the outside and it slides off whatever it hits. The other is angular and does a lot of damage and that's what we got hit by. Elverson was on the news during the storm—70 MPH winds and hail. We just got pounded! Only the hill Chard got pounded. I hesitate to guess on the loss, maybe 25% of the crop. The good news is that it isn't disease, and the berry damage is over.

Friday, July 21: I'm not feeling a strong pull from the past anymore.

July 24: The Japanese beetles are back already. They're in my hair and down my shirt, and hanging on my pants when I come inside from the vineyard. Meanwhile, the hail damage is drying up as predicted. Things are looking up.

July 25: Over the years, our lovely Vidal vines have gotten weaker, but still produce a full crop. I work with the vines to select shoots and position them on the trellis by hand. This morning Juventino appears in the vineyard and says he wants to start working again.

"You're suspended for two weeks," I tell him.

"I'm sorry," he says, "but that day I wanted to go visit my friend in Norristown."

"I told you the Friday before that I needed you here that Monday for the bottle delivery, and it was very important," I remind him.

He says, "Well, it worked out ok without me, so can I start working now?"

"No," I say, still working in the Vidal.

Then he says, "You can't get this work done without me, and if I can't get back to work now, I'm going back to Mexico.".

"I can do the work myself, so please leave now!"

August 1: We keep working in the main vineyard and get ready to bottle the white wines. The tanks must be emptied before harvest. My brain goes first to the grapes that are ripening. Have we done everything to help the vines? We finish getting ready to spray again tomorrow. More beetles arrive—they're eating the leaves from the top of the shoots down. I can hear them munching away all night long. In the afternoon, we start bottling the 2004 Chardonnay.

August 2: Miguel starts spraying, and Tulio and Gerardo work with me to finish bottling the 2004 Chardonnay.

August 3: The beetles are waning. We work in the

vineyard all day. After everyone leaves, I make the champagne starter for our first secondary fermentation bottling of the 2005 Blanc de Blanc vintage. It takes three days to make the yeast starter.

August 4: Each day, I work on the champagne starter, and then out to the vineyard in the morning. It's been sweltering hot every day, breaking records. I am busy on the Boston College project in my home office without AC. After lunch, we bottle the Gewurztraminer. It's a relief to be in the AC.

August 5: Preparations are complete to start the champagne bottling. I always feel festive on this day. Sparkling wine celebrates itself from the moment the yeast starter is added for the secondary fermentation.

Things go as planned. We finish 350 gallons of the 2005 Blanc de Blancs. Now we wait four or five years!

August 6: The weather breaks with a front that came through last night. It's in the 80s today with low humidity. I spend the day in the vineyard. It's glorious!

August 8: All day we prepare the vineyard for spraying.

August 9: The last spray is completed before the nets go up.

August 10: We start putting the nets up. The champagne grapes are first to net, and first to harvest.

August 16: Time is flying.

I take three days to make the next champagne starter, and always enjoy listening for the yeast working. Finally, we rack, filter, and bottle. The days are flying. Harvest feels like it is rushing in early.

The days are glorious, 85°-65°, with bright sun, and the grapes are soaking up the sunlight. We put up nets, one at a time; 17 feet up and over the trellis, 250 feet long rows.

August 19: It's been a rush to get ready for the meeting in Erie on Monday. This is the first Pennsylvania Quality Alliance (PQA) statewide members meeting with a tasting panel of Pennsylvania Chardonnay. It looks like it's going to be a great meeting. Guests are coming from the Lake Erie Wine Quality Alliance (LEWQA), and a private dining room at Mercyhurst College is reserved for the tasting panel. Two panelists from Delaware will spend the day with us, and lots of regional Vinifera wine tastings, and discussions on quality wines are planned. Finally, I think I hit the mark for our group for this meeting. It's taken hundreds of telephone calls to organize.

August 20: I can't think of anything else to prepare for tomorrow's PQA meeting. I meet Joann at the Coventry Mall parking lot to pick up her '03 Merlot for the tasting tomorrow. It turns out she went to my high school. She's about ten years younger, and her dad went to Drexel, like me.

At home, I decide to start mowing the severe hill in the new vineyard. It's a beautiful day with low humidity and clear skies. The barn swallows instantly appear and swoop around me, and very low and close to the Kubota mower. They come closer than ever before: they are taking risks, but confident on my driving. They surround me with acrobatics, swooping near my head and cross right in front of where I sit on the mower. They fly with me for an hour, catching whatever bugs are scared up by the mowing. I am propelled by their energy. There are six of them from the family that nested several years ago, and every year since. They have moved recently to the barn swallow gathering place near the lower marsh lands in the valley.

I discovered this one day while riding my bike through the valley. It's very buggy; a good place for bug-eating birds like the barn swallows. Two barn swallows flew close to me on my bike and followed me home. They come back every day when I mow, since it's good eating. They are visiting more now. I can tell they're getting ready to leave for the long journey ahead.

August 21: I drive to Erie. The meeting is from 1PM to 7PM, then dinner. We taste and discuss Chard's from around the state. We discuss the possibility of working on identifying wine districts for our members. The panelists from Delaware tell us why our program is important to them; it gives them something to talk about, and confidence to sell wines with PQA stickers. This is music to our ears. That's what we intended. Sometimes, one has to go outside of one's own area to get recognition. What a great day!

August 22: I visit Presque Isle Wine Cellars for the first time. I have bought equipment and supplies from them for over twenty years. Before leaving the area, I drive the vineyards near Lake Erie.

August 24: I ride, ride, and ride some more, with barely enough time to prepare for the Lance Armstrong Foundation (LAF) bike ride on September 10th. My plan is to ride twenty miles a day, and a fifty-mile ride on Sunday mornings.

I ask our customers for donations for the LAF bike ride. I will ride with Fred's name on the back of my jersey, and the customers are invited to sign the jersey before the race. About 50 customers stop by the winery to donate, sign the jersey, and wish me well.

August 27: I start the barrel *racking* and washing this week. All the red wines will be pumped out of the barrel and the sediment left behind in the barrel (the process called *racking*). We have two vintages of red wine in barrels at any time. The red wines that were in the barrel for two years will be bottled. All barrels are cleaned and sterilized.

I sample each barrel for taste and use that information to decide which barrels to use for the one-year-old vintage, and which ones to use for the new harvest. In addition, I use this information for blending trials. Long ago, we learned that the French oak barrels were superior to all others, and that new oak barrels were too strong for our style. So, a mix of older barrels was always preferred.

August 28: We had 3 plus inches of rain. The grape berries are holding–please no more rain!

It's raining in the old winery room. The stucco removal above the roof has damaged the weather seal. I'll have to get back up there to caulk. The winery basement has two leaks; from the side where the electric comes in and under the front door.

August 30: We bottle Demi-Sec this morning. After lunch, we finish racking the barrels; the 2004 and 2005 reds. The '05s are amazing!

We discover another leak coming from the sagging gutter in front of the retail room. We get the extension ladder and work to clear the copper gutter; there's small stucco pieces from the roofing project. Miguel helps to hold the gutter in place while I tighten the screws on the gutter hangers and reposition the gutter to be straight. This should hold for the next rain.

August 31: Just before last light yesterday, a bird of prey

silently arrived on the shed roof, about thirty feet away from the porch where I was sitting with Stars. Stars was eating some cheese and crackers, as dinner was hours away, and her back was to the shed. She never saw or heard the silent raptor. The raptor took off fast, swooping across the front lawn and behind the apple tree. I looked harder, trying to see through the tree, and then the bird appeared still swooping below the tree toward the new vines.

I check my Audubon book: it was a falcon—either a Merlin or Peregrine Falcon.

September 3: I ride with Patty and Joe in Lancaster, 45 miles. I wonder how my body will do on the century ride next Sunday. I barely have enough miles in; will have to rely on muscle memory. Hydrocodone is needed at bedtime.

September 5: Today is Mom's 86th birthday. I ordered fresh scallops from Truro last week, and flowers today. She is delighted. We're busy all day with more bird netting, tying up the little vines, and cleaning barrels.

September 6: All day long we do barrel work and get ready for the next bottle delivery tomorrow. The supplier swears we won't have any fly-by-night driver this time.

September 7: The bottle delivery prep is complete, and all of the reds are racked and blended, ready for bottling. When will we start harvest? Nighttime scouting in the vineyard reveals varmints are eating the Gewurztraminer.

September 8: More of the bird netting in the Vidal is finished. Now we mow, trim, net, and add extra perimeter nets to keep the varmints at bay.

September 9: The day before the Lance Armstrong Foundation ride in Philadelphia I am hurting all over. Stars brought me a flicker this morning. Peggy stopped by to sign the shirt I will wear tomorrow. She says flicker is for speed, fast riding. Stars must have known. Peggy spots a monarch chrysalis attached to the log cabin! Later, I find a flicker feather on the front walk. I will attach it to my shirt for the ride tomorrow.

Joe and Patty arrive for dinner and are sleeping over, since we get up at 4:30 a.m. They decide the flicker feather should go on my helmet and use packing tape to fasten it onto my helmet. We pack our cars, have a quiet dinner, and then we all go out for a vineyard scout, and to turn the bird machine off. Stars impresses them with her speed. In the dark, you can hear her thundering down upon you, and then in a blink she is past and out of site. They can't believe how fast and adept she is at navigating at such speed. Visions of riding are in my head.

September 10: I awaken at 2:00 a.m. and check and recheck that the alarm is really set, then doze off till 4:00

a.m. I take a hot shower, and then put the neuro-stim machine on. I feel okay and wake Patty and Joe. We have coffee and oatmeal with bananas and honey and gather our gear. I pack PB, crackers, and water. Stars waits her turn patiently, and then we run up the hill to turn on the bird scare machine. We run fast, but it's never fast enough for Stars.

We caravan out of the driveway a little past 5:30 a.m., and head east on Route 23, all the way through Valley Forge Park and onto Route 202 North. A few miles north of Norristown, we are in a long line of traffic. The car in front has a Maine license plate. I can see a bike inside. As we slowly approach the intersection of Route 202 and Germantown Pike, we see that we are in a slow-moving endless line of cars with bikes, inching to our starting point at Montgomery County Community College. It's after 6:30 a.m. and the starting time of 7:00 a.m. quickly approaches. Riders are abandoning their cars in every parking area off the road to make the 7:00 a.m. start.

Lance Armstrong will address the crowd before the start. This is one of six national Live Strong rides around the country. Finally, we arrive at the overflow parking area and get our bikes ready. I change into the special shirt that everyone has signed. Dad had sketched out six faces on a piece of paper and finally made one on the shirt. There are lots of hearts, good wishes, and signatures. The shirt feels light as air. Stars' paw print is on my shoulder and on the front, over my solar plexus. On the back of the shirt in calligraphy, in dark green letters with a gold edge is:

FRED

4/9/49 - 3/7/04

We hear Lance addressing the crowd, but we are not close enough to get a visual. The 100-mile start takes off, and then the 70-milers. We finally take off at the end of the riders and enjoy the ride without a pack around us. Hilly is the mode for the day, up and down, up and down, down to the river and then up the ridge again. The only flat spot is on the west and east river drives around the art museum.

I stop and rest at mile 46. We join the back of a paceline with some 100-milers and the Vermont team and navigate through the Sunday crowd down the east river drive. Then comes the climb up the hills of East Falls. We're in low gears and going slow. Patty and Joe have to stop frequently to wait for me. I apologize for my poor conditioning. They are gracious and cheer me on. We stop at the 55-mile rest stop. I am always dehydrated a little by this time and carry extra water. Next are more hills. Finally, we pass Chestnut Hill College and turn north. We have just ten miles to go. A nagging inner voice starts complaining: Wouldn't it be nice to stop? Maybe I can get a ride? Suck it up. I'm tough, right? Go, Janet Go. My head hangs down a lot as I press on. My chain slips, and I yell to Patty that I have to stop again. The road is rough, and my teeth chatter going downhill. Then uphill again, low gear, and one push at a time. My knee starts hurting and I try to move my foot a little, but it's locked in the cleat. I have to talk to myself to keep going. Several times Patty and Joe are stopped at a traffic light, and as I near, they say, "keep going, it's turning green." And so, we go for the last five miles. As I pass one police officer, and an emergency vehicle he asks if I have cramps. "No" I say as I keep my pace, and he replies as I slowly move past, "You got a big hill coming up." "Thanks," I say. I look down and my legs are still moving. Don't stop now. Suddenly it's one mile to the finish. I try to look a little less haggard, must look good for the finish. One more little hill, around the curve, and there's the last cross street. "Pick it up a little," I say to myself, as a 100-mile rider blazes past. I catch up to Patty and Joe and we ride to the finish. "Survivors on the left," the MC says, "All others on the right." I barely hear anything, he forgets to read my number and then says "oh, she has a feather in her cap!"

All day we said thank you to the supporters on the road, the volunteers at the big intersections, and the police with big smiles for us all day long. The route was marked with clear signs and reminder signs to stretch, wear sunscreen, and drink water every 15 minutes. The rest stops had signs with inspirational sayings. We have been surrounded in greatness to propel us to finish. So, we finish the race.

Thursday, September 15: It's been raining all morning. We're prepared to bottle the red wines anytime we can't work in the vineyard. It's easy to start and stop the bottling, because we don't filter the red wines. Sterile filtration is a must for white wines; in order to eliminate the chance a microorganism could cause some spoilage. Filtering can strip out some flavors in red wines, so premium red wines are often not filtered.

Friday: The rain continues and so does bottling.

Saturday, September 16: Taste of Harvest event, 2006. It's still raining but expected to stop by 1PM and sun by 2:00 p.m. This is the year of stalled fronts.

This new Taste of Harvest event was a big hit. I took customers into the vineyard to taste six different grape varieties, and then taste the finished wine. This is probably the first time anyone had this experience.

After dinner, Mom and I walk up to turn the bird machine off. I shine the flashlight up and down the rows and Stars gallops around. That doe is lying down again in the Cabernet Sauvignon. She just sits there and looks at me as I yell and run toward her. Finally, she gets up and Stars comes tearing around the corner after her. In a minute, they are both down in the lower vineyard. Back at the house, we hear Stars barking methodically. After five minutes, I decide to go investigate. She's in the lower Gewurztraminer by the road. I check all the rows with the flashlight on my way and call to her that I am coming. She has a raccoon cornered on the vineyard top of the trellis, under nets. I kick myself for not bringing the gun and tell her to stay, and I'll be right back. Stars starts to follow me, and I say to go back, and she does. I run to the house, grab the .22, some shot, and hurry back. I load the gun and then balance the flashlight in my left hand, and the .22 in my right, with a finger on the trigger. I'm one row away and lean the gun on the top of this row's trellis. It's an easy shot; I hit him in the chest, once and then twice. He rolls over. Mom says, "What happened?" I say, "I shot him." "You're kidding," she says. "Nope," I reply. "Well, he won't be eating any more grapes," she says. "Nope," I reply.

Monday, September 18: Based on the calendar, we

should have already picked the Gewurztraminer and the Chard for Champagne. I checked the Brix yesterday and all varieties are holding well. Miraculously, they seem to have ignored the last rain. Downey mildew has taken over and we're losing the canopy. Downey mildew is a leaf mold that will destroy the leaves on the vine shoots (called the canopy). All the vineyard rows are under nets by now, and it's hard to spray with the nets on. We should have sprayed two weeks ago, even in the rain. Now we have to spray, but maybe it's too late. Lots of leaves have fallen in the last two days, when it was still raining. We work all day to tuck the nets in close to the trellis, so the tractor won't pull the nets. We get the vineyard sprayed and some trimming done. Now we wait.

Tuesday–We bottle the Merlot today after more blending trials. It seems to have lost some of the herbaceousness of the last blending trials. The extra racking has improved it. We missed the spring racking. There's too much to do.

Wednesday: We start the bottling in the morning, while I work on the BC (Boston College) project deadline. When I check late morning, there are more bottles breaking at the corker. I look at the broken bottles and determine it's a cavity #5 problem. We are familiar with cavity problems from the champagne problem that took us almost a year to figure out, with no help from the manufacturer. We quickly isolate the #5 bottles. I clean the corker jaws during lunch, so we can finish bottling today.

September 17: This week will be very busy. We're working feverishly to finish the preparations for the BC project training session. Barbara called last night, very upset. The consultant brought in to do the session isn't getting it. He's not listening, and repeating errors. He doesn't understand the business processes. I spend several hours reviewing his work last night and devise a scenario where we take over the training session. Specifically, Barbara takes the lead. She's ready now. She couldn't have taken that on two months ago. We have to talk with Angela and the Project Executive and get it worked out today. Three days before the session with much to do. Harvest will probably finally start this week; almost two weeks late.

Wednesday, September 18: I'm frantic to get everything done. I will be at the BC training session on Thursday and Friday in Philadelphia. I'll have to leave everything laid out for two days. On Thursday, they must start picking at the top of the hill, where the highest sugars are. This will help reduce the pressure from the varmints. The grapes look fantastic.

At 10:00 p.m. I check the vineyard, shining the flashlight

up and down every row. A fawn is sleeping in the Cabernet Sauvignon again. I chase her down the row and Stars takes over. Then I turn the bird machine off. Nothing more I can do today.

September 19-20: I'm in center city Philadelphia for the BC training session. People are everywhere; not like when I lived there in the 70s, when it was dead after 6:00 p.m.

At home Friday, I'm relieved that everything looks good. The wine must is 19.5 brix: perfect for Blanc de Blancs.

October 1: Finally, we're into the main crush—Chardonnay table grapes. Every day, we keep up the vigilance against the deer and the birds. At least four more weeks of fighting for the crop. We're three weeks later in the calendar and face the mid-October frost date that shuts down ripening.

Saturday, October 14: Thankfully, there's been no hard freeze yet. Miguel comes to work this morning to help me in the winery. We rack the newly pressed wines into clean containers. After he leaves, I will spend the weekend cleaning the used containers.

October 16: Pick and press grapes every day this week and get ready for Champagne Day next Saturday. The grounds are groomed as best we can; wine bottles have been washed and labelled each day.

October 19: I must try to do everything possible to make the best wine.

- pick the grapes at the best possible time for wine
- pick out the rot by hand in the vineyard
- crush and press the white grapes right away
- use the right yeast, and not too much
- watch the temperature, all the time
- open the doors at night to cool the winery
- check the tops of the tanks, always, all day long
- don't allow fruit flies any contact with the new wine
- clean up all equipment and workspaces right away
- wash everything a second time
- smell all the vessels before use, and fresh rinse with a citric acid solution before using
- wash the hoses with the right solutions in the right order
- clean the containers right away
- be the sorter at the crush machine, so no leaves, stems, or rot gets by
- punch down the cap twice a day, one hour per tank, every day
- watch for rain
- patrol the vineyard at night, and early morning
- turn the bird machine on at daybreak, and off at night

The varmints are closing in and nibbling on the end vines. I have to get more press screen and tape the holes in the must pump hose. The starlings are flocking; must move the nets to the exposed rows. I have to keep checking every day, everything. What have I forgotten?

October 20: The day before Champagne Day we have everything ready for the event. The Merlot will be pressed and pumped today into barrels, and the equipment cleaned before end of day.

Lynette calls to say, "Please come back!"

I had left three minutes ago to pick up some supplies.

"Miguel had an accident. The press blew up and he thinks his arm is broken."

"What!!!!!" I turn the car around thinking that the workers were having lunch when I left: how could this happen?

I rush into the winery and Miguel is sitting in the labeling area with ice packs on his arm. Fortunately, Marsha, a nurse, is here today. I survey the injuries; a puncture type wound on the forearm is bleeding profusely. He says he thinks his arm is broken. Marsha says she doesn't think it's broken, because he can move his arm, and his fingers. I go around to the press pad and survey the remains of the press. It's blown apart. Parts of the press are everywhere as well as one-half ton of mashed red grapes and juice spewed all over the press deck, equipment, and the building.

It's a one-half ton capacity wooden basket bladder press. The bladder surrounds the center vertical rod, which is 2-inch-thick steel. The wooden baskets have three bands of three- inch by 5/8-inch-thick steel bands that reinforce the baskets, and the two basket halves are secured in place by six one-inch steel rods that lock the halves together. The press has a safety valve set to four atmospheres pressure. Now what the heck happened?

I go back to check on Miguel. This couldn't have happened at a worse time. The day before Champagne Day and we're already flat out, and in the middle of pressing the Merlot. It's Friday and the pressing must be finished. I ask Miguel how he's feeling. He answers by showing me his bruises on his leg, and around his middle, and his arm hurts. Everyone is flustered. We work on getting the bleeding to stop, but it doesn't. Marsha says she thinks we should take him to the doctors. She's changed her opinion partly because Miguel keeps saying it's broken, and he can feel it moving. My mind is screaming: an undocumented worker without health insurance. I go to the house to get some clean cloths and call my GP. "Are there any free clinics around?" and explain the situation. They tell me about the new Urgent Care in Elverson, which is just down the street. I call them and explain the situation. I say I will pay for the x-ray and the doctor's visit. They say ok, just bring him down. I help Miguel change into clean dry clothes and wash his shoes off. I tell Marsha about the local Urgent Care, and she is relieved. I get Tulio and Gerardo working on cleaning up, and I get the other press pulled out to finish pressing. They can get set up while I take Miguel to the doctors. Mom has arrived and will go to the doctors with me. Mom, also a nurse, follows me in her car. I remind Miguel on the way that I will tell them he is not an employee: I have no employees or insurance. He says ok.

I sign him in, and Mom sits beside Miguel in the waiting area. I explain about no insurance, but I will pay for this visit. They nod okay. I check with Mom and Miguel one more time before leaving.

Back home, I work with Tulio and Gerardo to get set up. They can stay late; we must get the Merlot pressed. I'm trying to visualize when Fred and I did everything ourselves. Since Fred became ill, I did what I had to do, which meant letting others do many things I would never have let anyone else do.

So, we pressed, with no problems.

Steve returned from taking the last of the ice wine grapes to the freezer in Lancaster County. He looked at the broken press baskets and said that there was a lot of force needed to break that apart. I had asked Miguel if he left the air on. That had to be the reason for the "explosion." Tulio had told me that he thought Miguel had left the air on a little bit when he went to lunch. Miguel had said no when I asked, but there is no other explanation.

I remember times when I have caught Miguel doing something dangerous to cut corners and taking unnecessary risks. It would follow that he would leave the air on a little bit during lunch to hurry the work along. He must have done this before. It didn't work this time. Finally, we finish pressing the Merlot. Steve has stayed to help. We are all on maximum overdrive getting ready for tomorrow and the 300 guests. I must rack the Merlot in the morning and fill barrels, then finish cleaning the winery before anyone arrives. I do everything I can till dark.

Meanwhile, Miguel had a compound fracture and will need surgery immediately. The orders were written, no food or water after midnight and surgery will be in the morning.

October 21: Champagne Day—I check outside at 11AM and customers are waiting by the entrance already. Champagne making tours are to take place on the hour. Tasting the new champagnes will be in the courtyard. Complete wine tastings are in the tasting room, and a band is to play outside. It's a beautiful day, and very good sales.

October 28: Miguel arrives with friends, with his arm in a cast. Miguel says that he can't work for four months and wants me to pay him full pay for four months. I say no, and that I made no commitment for him to work this winter. Ordinarily, he would stop working after Thanksgiving, and start next year in April.

November 28: Miguel arrives, and his cast is off. I tell him my final decision is to pay him for full time for 8 weeks. That would be his regular pay if he worked until mid-December. He says he can't work for four months. I look at him, and say your cast is off; you are right-handed, that arm was not injured, and you have two legs. You can start work on Monday for full pay, and work for the winter. He then agrees to accept eight weeks full pay. I say that we will sign an agreement and hold out the clipboard for him to sign. After a second, he takes the clipboard with his "injured" arm and signs.

December 3: A northern Italian food and wine tasting dinner at St Peters Inn was my Christmas present to "the crew." The old inn has finally been restored to prior grandeur. They're selling our Blanc de Blancs for $60/bottle. I hope they stay in business.

December 6: It occurs to me while swimming that maybe I am grieving the loss of myself, or is it the lost period of time? I said to Mom, I'm still grieving from the horrible and shocking experience of losing my husband at such a young age.

December 9: Stars has been limping all fall and finally I take her to the vet. She has the equivalent of a torn ACL. I carry her around in and out of the car.

December 15: I'm up at 5:00 a.m. to pick up the frozen grapes. Today, we start the arduous task of pressing the frozen grapes.

December 20: Day 6 of pressing frozen grapes. I start visualizing the last day.

December 23: Finally, today is the last trip to the freezer. Back at the winery, we clean the presses then load the grapes. I can make it today, but no more.

December 31: The winery is open till 5:00 p.m. sales have been slow this week. The accounts payable are current, with a cash infusion from the retirement savings again. Since Fred's passing, I have used many extra hands to put in the new vineyard and help with the extra work. I'll need to be more efficient next year.

Successful small wineries need to be at least 2000 case production. That's 24,000 bottles of various shapes. Just a few of the costs are:

- $1 per bottle
- $1 per cork and closure

- $1 per filter, yeasts, and supplements
- $1 per label

That's $100,000 per year sunk cost before putting wine in the bottle. Wine production costs include growing the grapes, vineyard and winery labor, equipment, supplies, buildings, maintenance, insurance, utilities, and taxes.

It's a constant struggle, just to do it all again the next year.

So why do I continue? As I said earlier, I like to make things, and I like to grow things. If you ever had the chance to grow the most incredible Chardonnay vines and taste the ripe grapes directly from the vine, you would never forget it. You would do whatever is needed to care for them and nurture them, so you would get to taste them again and again. You would become in harmony with them, and the grapes will give you a bounty suited for the gods.

2007

January: This is the resting month of the year. I spend it feverishly working on cleaning and refreshing my office room. I scrape, sand, and repaint the old caste iron radiator, replace the baseboard, and empty the room of mountains of paper. Then I repair and paint the walls, the windows, and the window hardware. I am now ready to take on 2007.

February 1: Pruning starts the first of the month. I'll have to make do with only one worker. I need to check again with the Township on the sale of my development rights. Those monies would cover the cost of the second mortgage, freeing up monies for another worker.

February 3: Finally, I take Stars to the visiting orthopedic specialist. After 1 minute, he says that she needs surgery. He's done 500 of these surgeries. The options are be referred to another specialist in the area, wait 5 weeks for this specialist to return, or go to Wilkes Barre to his new office. I call his office immediately and get scheduled for Monday. He gives me an overview of the recovery—six weeks, no furniture, no stairs, four weeks "bed rest." We'll sleep in the living room and block the stairs off. We'll both miss our bed.

February 14: We're snowed in with another six inches. I saw another eHarmony commercial and decided to investigate. After taking the personality profile, I get seven matches right away.

February 19: We (Stars and me) get up early to give Stars a couple of walks before our trip north; a long ride, and the turnpike is still not all cleared from last week's snowstorm. We arrive around 11:00 a.m. to a new facility. No one called to confirm, so I call the doctor on the way up. He calls right back, and says yes, he's expecting Stars. After checking in, Stars knows something's up and climbs up into my lap and tries to climb out the window. Back in the car, I ask my angels to stay with her, my best buddy.

The rest of February is a rush of pruning and taking care of Stars. I set up the foam mattresses in the living room again, and we sleep on them. I hold her next to me under my arm, so she doesn't have to wear the collar at night. I talk to her softly about how this is only temporary, and after a few days she'll feel better. Outside, she's a tough walk on a leash, strong enough to pull me up a hill. I am determined to do my best to walk her, so the repair takes.

eHarmony.com gives you *matches* with their questions and answers, and the opportunity to send your own questions. After some back and forth, I schedule a telephone meeting with Jamie. I ask if he is under the care of a medical doctor, and he shares his recent recovery from cancer.

March 21: Jamie and I meet in the Park City Mall. It's awkward for me, and I have much trepidation. We sit in the center court and Jamie talks about his clip-on sunglass mall business.

April 26: Jamie and I talk a lot on the phone. He is a good talker, and I like to just listen. He is recovered from stage-3 colon-rectal cancer. People ask, "How can you just start a relationship with someone you really don't know?" I reply, "Because cancer survivors don't bullshit."

May: Bud break continues with gorgeous weather. The new green shoots decorate the landscape. We finish removing the dead vines and planting the new ones. Modest rains continue through the spring.

My commitment to continue the vineyard was made last year, with the new vineyard land purchase. Now, I must determine how to proceed forward.

Automation will provide efficiencies, but the costs are high. With a new Cab tractor and bigger sprayer, I can do the vineyard spraying myself in three days. Also, a forklift attachment will be a major assist and savings in unloading bottles and other deliveries. In the long term, this is a good decision, but requires more cash out of the retirement account.

There are new narrow tractors designed for vineyard and orchard applications. Rich Blair offers to transport his new narrow tractor for me to try. Once on my hill, it is obvious that this is too tipsy for my land.

June: Ideal weather continues with full sun and weekly half inch rains. There is no disease pressure anywhere in the vineyard. All young vines are growing well, and we manage to keep the shoots tied off the ground, lest they are run over by the tractor.

The new Ford New Holland tractor arrives, and I get a full checkout on the attachments and the sprayer. I start immediately and go right to the top of the hill determined to overcome any trepidation. I quickly learn how to keep track of the row pattern, so you can keep the tractor moving forward as you rotate through the rows. The high rpms of the PTO on this tractor make me slightly nauseous. I'll have to rest in between spray days.

August: Jamie comes to visit one Saturday in early August. I am working on preparing the vines for the last spray before netting. He sits on the house porch with his laptop and notebook for the afternoon, while the winery is open. The next day we continue the same; he on the porch, and me in the vineyard, while the winery is open. When I close the winery on Sunday night, I always settle the cash drawer and the credit card receipts. Jamie asks me what the wine bottle sales were for that day. When I conveyed what the sales book said, he replied, "That's curious because I saw twice that number of bottles carried out the door." We compared notes for Saturday and the result was the same.

The next weekend when Jamie was visiting, we conducted a physical bottled wine inventory. When Fred became ill there wasn't time for everything, and this was one of the items that was left undone. The result of the physical inventory was that 200 cases were unaccounted for. I remember that one day, when I was coming home from getting hospital supplies for Fred, Robear, the tasting room employee, was in the driveway at the back winery door loading cases into his car. At the time we were only open on weekends, and Robear worked both weekend days. He also had keys to the winery, since we were at the hospital long days, and needed a few people to be able to get into the house or winery, if needed. He said the wine was for a benefit that Fred approved.

Over the next three weekends Jamie did "porch" duty to keep track of customers and wine bottle sales. The cash and credit card receipts always totaled the registered bottle sales but didn't match the physical inventory going out the door.

We installed a small video camera in the office above the tasting area. We learned that Robear was pocketing cash for wine bottle sales and selling glasses of wine for cash.

The next weekend I confronted Robear when he arrived at the winery and said, "We know what you've been doing."

"What do you mean?" he innocently replied, as he went through the motions of setting up the tasting area.

"We installed cameras," I said. He looked down and paused. "Leave now," I added, glaring at him. He turned to collect his things, and at the door said, "I never meant to hurt you."

Harvest begins after Labor Day with the Chardonnay for champagne. The grapes are in perfect shape, and the weather is dry. We enjoy beautiful days of picking the grapes in the morning and pressing in the afternoon. After two days of the juice settling inside, we rack into clean containers and start the fermentation.

One day in mid-September, on my first walk around the vineyard, I become aware of a humming sensation as I walk the length of the first planted Chardonnay vines. I look around for an electric machine or something that would cause a vibration, but nothing was in sight. Later I return to this row, and again experience the humming sensation. I conclude that the vines are humming. I have no proof, only my own senses.

The grape berries are small, and the yield is 30% less. The Gewurztraminer has never shown this much varietal character. Not many growers try Gewurztraminer; it's known as "finicky" in the vineyard and winery. Another Chester County grower calls to ask if I have harvested the Gewurztraminer yet. "No," I reply. "When?" he asks, saying he has 20 Brix and the critters are all over it. "It's not yet optimum ripeness. You just have to wait, and protect it", I reply. We waited ten more days and were rewarded with a special vintage. Years later, customers came back asking for it; they also knew it was special.

October came and went without a hard freeze, The red grapes were still in good condition, and we waited till mid-November to harvest them. The color at press was intensely dark.

This year's Champagne Day was our first ticketed event. Up till now, our events were free to customers, and by invitation. The events had grown in attendance by word of mouth and were starting to resemble tailgate parties. This year we set up a check-in booth and required a rsvp, or their invitation card. The fee was $15, and attendees were given a logo champagne flute, tickets for ten wine tastings, five champagne samples and appetizers. The attendance was one third the prior year, but the same bottle sales. A few customers wrote letters complaining about the fee, since they always enjoyed bringing their friends to the event in the past.

When Fred became ill, and for the years since then, I needed extra help. The solar grant project had ended, and I had no time to find more consulting work for myself and Lynette. Marsha would be needed through Christmas, but in the New Year, I knew I couldn't afford to keep them.

Jamie and I had continued our relationship and were discussing his interest in helping me with the winery business. He had a lot of business experience and thought that our Vinalies award should bring the winery to the next level with additional investments in advertising.

After the incident last summer of the dangerous nighttime visitor, I began work on designing a gated entrance. Randy, the stonemason, built two walls this year on either side of the driveway to frame the entrance. Philadelphia has many beautiful churches with wrought iron fences and gates, and I did reconnaissance there for ideas. I discovered Philadelphia's first blacksmith shop in Powelton Village was still operating. The owner, Matt Weber, was in the shop when I stopped by to discuss my design ideas, including pictures of the Christ Church gate. He excitedly went for an old book that had drawings of that gate. He agreed to visit, and after careful discussions he agreed that some of the pieces would be made in Austria, since there aren't enough blacksmiths in Philadelphia to do the complete job in a reasonable time.

2008

January 6: I order the Belgian block for the driveway entrance; five tons of jumbo. This will cover the 30-foot area from the road to the gate and continue inside the gate another 30 feet. The entrance is 22 feet wide.

January 8: The block arrives, and Paul starts laying out the work. It takes all day to set up.

January 10: The mild weather from New Year's is gone and replaced by sleet and hail. We start to lay the block. Paul is the leader, myself, Tulio, and Armando follow. This is very hard work since the blocks weigh about thirty pounds. Each one is placed after much sorting and surface prep. It goes painfully slow.

January 15: Every day it rains, snows, sleets, or hails on us, as we lay one block at a time. We have only six rows done. My back is hurting. The wind is howling.

January 19: The days go by in a blurry of getting ready for the next day, with stone deliveries, running for supplies and tools and working each day with inclement weather.

January 20: Paul has heart palpitations and is in the hospital. We have a wine trail event coming up in two weeks. The driveway has been closed since January 2nd. We have opened the field gate and laid down straw. Every day it must be groomed to keep it passable.

January 23: The credit card companies really socked-it-to-us in November and December. Our rate was pretty consistent all year at 2.9% but suddenly jumped to 8%. We will look for another company.

January 30: Jamie and I decide for him to move from Lancaster to the farm. He offers to help with the monthly costs and capital projects. He's very good with customers and takes the lead in the Tasting Room on weekends, freeing me up for the vineyard and winery work.

February: I try to prune every day, but most days I work only on the driveway project. The weather is miserable; cold and windy. We finish the Belgian block and decide to pave the entire parking area with bricks. We're striving for an elegant outside look to match our wine style.

Jamie offers the capital for the brick driveway. We visit Glen Gary in Reading and pick out Belgard pavers and decide on red stones for the vehicle parking area. Paul uses the tractors to demolish the parking area and gardens. He lays out twenty feet of work area for the pavers. The pavers start at the Belgian block and will continue for

160 feet in length and 22 feet wide. Armando, Jamie, and I work a little each day moving the pavers to the work area by hand, laying them in place, and hand tamping, one at a time. The days wear on, my shoulder hurts all the time, but my back is fine. I keep laying brick every day to keep the crew going. We mark out 20 feet with string line, then lay in screenings, tamp them, and then lay in pavers, one line at a time. Sunup to sundown is full of driveway work, prepping for the next day, getting supplies and pruning. Pruning is always on my mind; what row, which vine, how many more to go.

I travel to Harrisburg to meet with the PLCB to present our new champagne label, which was designed specifically for the PLCB stores, and to ask for an order. I took several samples of our new ice wine, which is always favorably received. They will get back to me when they confirm the spring orders. We do not get an order this year.

March: We're still laying brick but are almost finished. The field gate entrance is worn down and muddy. We spend time every day on this temporary driveway entrance trying to keep it passable. I manage to prune a lot on the weekends, with Jamie taking care of customers. The daylight is longer, so I get a couple of hours in the vineyard after everyone leaves for the day.

Robert, the Sommelier at St Peter's Inn, is now at

Flemings in Radnor, Pennsylvania. He organizes a Women in Wine dinner with the winemakers featured in Deborah Brenner's book "Women of the Vine." He invites me to be included and have my champagne represented.

April 1: The day is filled with more work on the driveway project, trying to get the new merchant servicer up without success, and finally, time for the vineyard in the last hour of the day. This is the second rain day in a row. I'm behind two more days; the race is on; the buds are activating.

I'm making good progress. It's a little before 5:00 p.m, when suddenly Stars gives her fierce growling bark and dashes into the rows below me. She is insistent that the intruders take heed.

It's a couple arriving for a late day visit to the winery. Stars has been watching intently for a few minutes, then goes into her resounding fierce bark the moment they set foot out of their car. She growls and barks and follows their path to the winery with her eyes, for we are up at the top of the vineyard. I say, "Go ahead down and check on Jamie." Fifteen minutes later, I hear Jamie saying goodbye to the customers. He runs up the hill and comes bounding down the vine row in greeting. We smile, thankful for each other's company.

April 2: I hitch up the old green van and cart, checking

to make sure the chains are in place in case the hitch fails. We don't have a pick-up truck, just a pair of old Dodge caravans.

Paul is working on a small retaining wall from our field stone in the parking area. I'm off to get the next load of bagged concrete. The tow hitch on the old green Dodge Caravan's rated for 2000 lbs.; that's a ton or 25 bags of sakrete mix. While inside the local hardware store talking to Jim about his latest paleontology find, the forklift driver attempts to load a full palette onto the cart. I arrive to find the cart up in the air with a broken hitch. "I wish you'd waited for me." I say, trying to think about how I'm going to get this fixed. Half of every day is spent fixing something. Jim, the amateur paleontologist, comes out with a sledgehammer, and we figure out how to pound back the stuck rod to remove it. It's not long before he comes up with a temporary bolt and I'm back on the road. Best place to break down: your local hardware store! I only lost an hour.

April 18: Finally, Paul is finished the driveway work after tamping the red stones. We help clean up his tools and pack the car.

After he leaves, Jamie asks, "How are we going to finish?" I explain that is how every job goes. We have to finish the rest ourselves, including:

grooming the dirt

removing stone screening piles on the grass and driveway

edging all the new gardens

adding more topsoil and grading the destroyed gardens

seeding, mulching, planting

grouting the brick and Belgian block

sealing the brick pavers

We have essentially recast the entire entrance and gardens. It will take months of working little by little. Jamie looks at me in disbelief. He doesn't know what I know I can do.

I continue pruning every day; each one gets a physical cut back, tied up, one by one, up and down on my knees for each vine. I'm sick with a cold-virus bug; sweating all day and all night, constantly wet, flushed, cold for over two weeks. I tell myself just prune a few rows each day, vine by vine, hour by hour.

April 19: Sunday, ahh, a day of rest. I go out for a walk around the vineyard mid- morning and within seconds Stars has something cornered in front of the kitchen. It's a snake; sandy and black markings, coiled like a rattler, small. It lunges for her as she gives it her best loud ferocious bark. She taunts it from one side and shifts quickly to the other side to try to catch it off guard. It follows her and lunges again. I call her to back away and run for the new digital camera. I can hardly believe that I am trying to get on the ground nearby the snake for a picture. I hold my breath that Stars will stay still and so will the snake. When I get up, Stars puts herself between me and the snake, to protect me. Finally, I back away and she deems it a safe distance and comes to my side. Later, I look up Chester County snake pictures—it was a copperhead.

The rest of the day is spent with more pruning, rain and thunder, more pruning, and a group at 4:00 p.m. in the winery.

May: During the first week, Matt Weber arrived with the gorgeous reproduction Christ Church double gates. He and his crew installed the spindles in the stone pillars and hung the two thirteen feet high and ten feet wide gates. The electric openers had been installed in March with the sensing circuits under the Belgian block. The gates opened majestically.

July 11: The barn swallows have fledged, and greeted me en masse as I stepped outside at 6:30 a.m.

July 12: In mid-day, I step outside the kitchen to see a baby barn swallow sitting on the brick walk; not a good thing to do with a bird dog around. I scold it into flying and it turns around and squawks back at me as it perches on the arch above the walk. I watch closely all day and see the mother feeding them; the parents are always close by.

We trim the grass edge at the driveway and sweep up. It's a scorcher, and we curse the heat.

July 13: I head up to the old Cabernet Sauvignon, determined this year to give it the best care as it looks good for the first time since 2004 when 80% of the vines died. I'm working my way down a row when something bites hard above my knee. I pull up the pant leg to see a red mark, but no cause is evident. One minute later I yelp as something bites me on the thigh and it stings like crazy. I have no choice—I drop my pants. I'm close to the road and look across the driveway to see if the neighbor is in view. Our road is an old narrow country road with little traffic. All clear, and I throw down my gloves and shears, untie the work apron, and finally yank my pants down around my ankles looking for the creature. I can't find anything. Then...when I get the pants turned inside out, and back outside in, there's the culprit—a lowly ant. Never again will I underestimate an ant.

July 15: Jamie comes over to meet me as I come in from the vineyard to get lunch. He describes an unannounced visit from the Penn State Worker Protection Specialist, Jim Harvey. He had emailed me last month about worker training, and I had relayed that we have no employees, and that we had a visit from the EPA last year. Well, he was in the area with a co-worker who supposedly had never visited a winery, and questioned Jamie at length.

"We're a four-person team; Janet and I and two hourly workers," Jamie told him.

"Right," he says, "And don't you get workers for pruning and harvest?" Again and again, he asked the same question, and said that he could supply workers. I didn't know that Penn State was in the labor business, or maybe he's trying to make his own business.

July 18: We push ourselves to do every little thing in prep for the upcoming photo and video shoot. Every square foot is as manicured as possible. We're so exhausted. We can't sleep.

July 19: At 5:00 a.m. Stars gives a soft bark. "They're here," Jamie says. "They're at the top of the vineyard waiting for the sunrise."

We assist the photographer and videographer in their activities; driving them around to suggest various shots. Steve shoots still shots and Allen shoots video. About 2:00 p.m. we're all out of steam and we have lunch in the tasting room together on Great Aunt Mary's 1865 table brought over from Germany, like a farm family after a long day's work.

July 21: This year I can't seem to get the spraying complete in one day. I'm starting later, spending more time in preparing the vineyard. The canopy is harder to manage with the frequent rains.

July 29: I'm 58, and not counting. No time for a celebration today since I have to spray. More rains occurred yesterday, and I am worried about the Downey mildew on the leaves.

July 30 – More vineyard work all day, then finish spraying after dark. I quickly rinse off the tractor. I'll have to do a thorough cleaning first thing tomorrow.

August 9-10: The Berks County Wine Trail (BCWT) summer event. I was invited to join, this past winter even though we're in Chester County.

August 11: The BCWT meeting starts without progress on the non-profit tax status. I offer to ask my new accountant, and everyone agrees with relief. Feedback from the recent event turns to questioning me about our wine tasting fee. A wine tasting of two champagnes, two white wines, four red wines and two ice wines was available complimentary with wine purchase, or for a $5 tasting fee. Several BCWT members said that they got complaints about our tasting fee. "I'm not giving my wine away for free anymore," I reply emphatically.

August 19: The Kubota is broken again. I clean the tractor,

tip the seat forward, write a note to the Kubota service man, and put the tractor in the driveway for pick up.

August 22: The customer who supposedly complained about us to the other BCWT members writes a long letter asking that we consider giving away free tastings for the wine trail event and says that Tom Calvaresi (a BCWT member) agrees with him.

We continue working on the bird netting and hoping that the Kubota is back soon.

As I walk across the front yard toward the kitchen door, a hummingbird flies past my head and actually brushes my cheek. Jamie saw the flyby, and marvels at these incredible creatures.

August 23: I resign from the Berks County Wine Trail.

From: Berks County Wine Trail
Sent: Tuesday, August 26, 2008 8:49 PM
Subject: Janet - wanted to share my thoughts ....

Dear Janet:

Please reconsider your resignation; this trail needs you—not just because you are a talented winemaker and blessed with the green thumb to coach endless acres of grapes to grow or that you are woman-owned business—but the simple and clear fact that you are bold.

Yes, bold.

Every wine trail should be so lucky to have someone bold like you on their team.

Why bold?

Because you boldly sent bottles of your wine to faraway Europe to win a fabulous prize.

Bold because you believe so much in what you do that you planned a dramatic expanse of vines.

Bold because you have the sense of humor to make slushies out of coveted Ice Wine.

Bold because you believed in a little trail with a diverse philosophy about winemaking and joined in on the sleigh ride that since gone from a walk to a spirited full trot.

Bold because you made such historic farm buildings vivid again.

And bold because you fervently said at our last meeting, "I'm not giving away any more of my wine," stated because you believe in the quality of your product and what you are doing as a winemaker and that others should have that same respect for your work and product as you do—and you took a stand by your beliefs.

This last act of boldness made even more bold—because you took this stand in a troubled economy and in a county known for its quirky, penny-pinching frugality—and where no one else was doing this in the immediate wine community.

When I was in LA, one woman would toast another woman who did something super bold by cheering, "Cajones!"

So, I am writing to ask that you please take a moment to re-consider your resignation because you are the bold.

Because you are the bold remaining part of the trail would make everyone involved a better person, including the misinformed individual who said something so carelessly. And isn't that really what we're all about—helping our fellow man become a better person as the penultimate in bold?

Regardless of what you decide, you are a terrific winemaker and a killer businesswoman and farmer. My hat's off to you, Janet.

Cajones!
Holly

August 27: In mid-afternoon Jamie comes to tell me the Kubota mower stopped working. I heard a loud noise last night, but I couldn't find anything wrong.

When I get back to the house, the UPS driver is walking down the driveway. The bird netting is here; a 17-foot roll weighing about 400 pounds. We muscle it off the UPS truck, and Tulio notices that it's torn. The UPS man asks if I want to refuse it. I'm desperate for the netting.

"Can we make a report that it's damaged?"

"Yes," he says and quickly asks about how much is damaged and the cost. I determine that out of 5000 feet maybe 1000 feet is damaged. He says he'll write it up. We quickly move it onto the cart behind the green van. I drive into the vineyard and drop the roll uphill a little. We can't do anything with it until the Kubota is back; it's too unwieldy to unroll by hand.

September 4: My arms are about to fall off after hauling the bird netting around. I change places with Armando and pull the newly cut bird netting out into the rows. One net at a time, about 240 feet, 17 feet wide, pulled carefully between the rows, then put up and over the row, and tucked at the ends to prevent birds and critters from getting at the grapes. It's 95 ° today. At 3:00 p.m., I finish putting the last net up for today. I pass out in the house in front of the computer, checking once again for the upcoming weather event. After more water, I grab ice pops for Jamie and me, and head over to the winery office.

This year, we will have all rows netted and are adding extra netting around the perimeter of the vineyard to block all entry into the vine rows. I tried this technique last year with success. For night patrols, I check the perimeter nets, and have more confidence that the grapes will be left alone. It turns out that deer don't like to walk on the nets, so I lay a few feet of netting on the ground around the perimeter.

Thirty minutes after the nighttime patrol, Stars is barking loudly and not stopping. I grab the .22, some shot, and the flashlight. She's in the Gewurztraminer. I load a bullet and head towards the vines, flashing the light down each row. Suddenly, I hear a fierce snarling growl that makes my spine tingle. It must be a raccoon. I call out for Stars to wait for me, then I hear another hissing growl, and see her jumping up into the trellis. I call out again as I scramble under the perimeter fence and drag the rifle carefully. I leave the safety on till the last moment. I realize that the vines are too high for a shot. Stars is jumping up at the critter, who is running back and forth on the top wire. I navigate under the two layers of bird netting and manage to crawl through with the rifle and flashlight. The .22 catches on the bird netting and I rip it away quickly watching for the raccoon. It's big and completely focused on Stars. I call to Stars to stay down, and shoot at the body, the raccoon falls and starts to run turning to hiss again at Stars. The .22 gets stuck in the netting, as I try to get a clear shot. Finally, I get the second shot, and watch for movement. I decide on a third shot in case the animal is still alive. Stars sniffs and sits down panting. She has had furious activity for 30 minutes keeping the animal "on point." I wait for her to recover. She tries to pick up the raccoon but gives up; too big to drag. She won't leave the catch, so I get the golf cart and go on patrol. She joins me after a time in the upper vineyard, and barrels downhill toward the house. She comes in shortly after I take off my shoes and flops, panting so hard. It takes her 20 minutes to cool down. Later, she is so tired she passes up cookies and milk and hits the sack.

September 6: Jamie is upset; nothing is in his control. Everything here is relentless and so hard all the time. I say, "That is how my life has been; maybe that's why I can endure it."

We just have to take it as it comes. There's always another obstacle. Just when it seems to calm down, something happens to remind you that you are not in control, and never will be. You better be able to float with it, or you won't make it.

Today was the hurricane effect. It's been raining torrentially all day. It makes me sick to think about the grapes.

September 7: Jamie takes off on the mower, and I take off in the golf cart to open up the netting in the rows in the old vineyard so we can finish mowing. Shortly, Jamie walks up and says the Kubota is broken down again. I try not to show any reaction.

"Okay, let's go see." We take the golf cart to the Kubota.

I try to run it and determine that the hydraulics are not working. We go back to the house to check the manual to see where the dip stick is. It's in the back under the seat wedged along the side of the transmission, only a child's hand would fit there. I wrestle it out, and yup, it's empty. Back at the shed, I search for something to use as a slide to get the oil into the tank, since only a 2-inch round object would fit anywhere near the tank. Finally, I think about a hose on one of the winery funnels. I cut off a section of hose and go back to the Kubota. The hose goes onto the funnel, and I get a gallon of oil into the transmission. I start the engines and wait for a minute for the oil to circulate. The hydraulics work. I drive the Kubota back to the driveway for pickup on Monday.

September 9: It's Tuesday and we're late on everything, trying to hold off buying bottles and supplies as long as possible. We are full to the gills everywhere and move a lot of things around to make space. It starts raining, so we abandon the vineyard. I test all the white wines for SO2, acid, taste, and check for clarity. They are all wonderful. We keep working to get to the bottling of the Viognier. I prefilter it into a clean container, and then set up the sterile filtration. We finish around 5:30 p.m.

September 10: Harvest begins. We pick the top eight rows of Chardonnay, picking out any rotten berries from the hail. Mostly, these berries are dried up, and easily fall off. The fruit looks beautiful. We store the grapes inside.

September 11: We pick 72 lugs (about 2000 lbs.) of Chardonnay in the morning. After lunch, we finish setting up the press deck equipment, and finally start THE CRUSH! I have a small group of Women in Agriculture visiting for a Taste of Harvest this afternoon. I take them into the vineyard and pick grapes from different varieties and taste the corresponding wine. I must ask them to watch us crush first, before the Taste of Harvest. They had an unbelievable experience. We finished pressing after the group left and were cleaned up by 5:30 p.m.–200 gallons of astonishingly glorious Chardonnay juice for champagne!

September 12: We pick Chardonnay through lunch and get 82 lugs. The drizzle starts before lunch, so we move the lugs to the press pad area and cover them. We'll hope for a break in the rain to press the grapes today.

We end up using tarps to create an awning over the covered grapes. All night, I worry about the grapes under cover, periodically getting up to make sure the tarps are holding. The next day we dry off the equipment and press these perfect grapes.

September 14: It's going to be 92 degrees today and very humid. After breakfast, we sweep the driveway clean from yesterday's trimming. I check the winery and wash the next load of bottles for labelling. I walk around the winery work rooms planning the best logistics for bottling in order to have the tanks empty in time for the next grapes. I weed the house garden and start pulling out that Burpee's "plant of the year" cosmos, with flowers you can't see, and huge seeds that stick into your clothing and hair. I'm always pulling them out of Stars. By noon, I am covered with these seeds sticking out everywhere. I look like a porcupine. When I get back to the house, there are some winery visitors walking around the vineyard with their dog. I head down the driveway to speak to them.

"I'm sorry, but you can't walk your dog in the vineyard. I have a dog here who guards the property."

"Ok," she replies, "I'll put the dog in the car. Can I get some water?"

There are visitors who just come to look, go to the bathroom, get some water, let their dogs and kids run everywhere. Later, we pick up paper, cigarette butts, the dog water bowl, and scrape chewing gum off the driveway.

Finally, the gates are closed, and I change for a bike ride. In the front yard is a barn swallow perched on the garden arch. I approach slowly to make sure it's one of ours. I've been thinking of them all week, knowing they will take off soon. He's come to say goodbye. I talk to him softly for ten minutes. I wish a successful flight, and hope for their return. He quietly sits, moving his head around, and occasionally glancing down at Stars. Stars looks at me while I talk; a tone usually reserved for her. She sits down beside me and waits. I hate to leave and say my last goodbye.

The wind blows strong, as the weather front is pushing through; from 92 degrees to 70 degrees in three hours. The barn swallows leave the area with a strong weather front; flying due east out over the Atlantic, then down the coast south, and then across land, as they continue south to Central America.

September 15: The bottle delivery is today. Yesterday, I couldn't get the loader on the new tractor due to inexperience. I tried for an hour and could only get the loader stuck onto the connection pins. I call Steve and he graciously agrees to come over around 6:30 a.m. with his equipment.

Steve arrives promptly with his tractor and parks near the field gate for the bottle unloading. Then we walk to where my tractor is, and he looks it over.

"I bet we can get this on," he says.

I call the trucking company since it's now past the delivery time. This company is usually very good and calls back in two minutes. The truck is on the Schulykill Expressway outside Philadelphia. Why do all truckers want to drive RT 95 into Philadelphia and the Schulykill Expressway out to the Pennsylvania turnpike? It's the longest possible way to get here, and still be driving in Pennsylvania. A four-hour trip from Geneva NY is now a seven-hour trip. Anyway, it will be 90 minutes or so till he arrives.

Steve and I work on getting the loader on the New Holland. He shows me a few things about positioning the tractor and it drops in! We get the bucket off and get the pallet forks on in a few minutes. I am grateful to Steve for taking the time and helping me with the loader. Tips mean a lot, and when you're a girl playing in a man's world, they don't come very often.

Steve waits with me until the truck arrives, and then breaks down his equipment while I start unloading. As I first approach the truck with the tractor, the driver waves me off "a woman driver?" he says. "Yup," my standard answer. I unloaded twelve pallets today, in less than an hour and a half. The trucker is holding the paperwork for me to sign and remarks, "I was worried, but that was pretty good work!"

September 22: Harvest is in full swing. We are working madly to pick, crush and press the whites. I prepare the Champagne starter and begin preparations for the secondary fermentation bottling, much later than usual.

The winery sales have been good so far and we are optimistic about a good fourth quarter; most important since it is 50% of the year total. We decide to be open seven days 10:00 a.m. to 5:00 p.m. in November through year end.

It's around 5:00 p.m. and we need the Kubota to empty the presses. Jamie has the Kubota for some mowing, and I wonder where he is. I say something to Tulio, and he says Jamie is in the house, but where is the tractor? I go down to the house and find Jamie just leaving the kitchen with scissors and meat cutter in hand. "Where's the Kubota?" I ask. "Don't ask," he responds. Baffled, I continue, "We need it." He retorts, "Don't talk to me." Again, I query, "What's wrong?" Finally, he spills out about how he was mowing, and all of a sudden, an entire row of bird netting got caught under the mower. He yells at me to stay away. I convince him to let me help. At the top of the hill, we find the Kubota stuck in a heap of bird netting. I start cutting the netting away from the mower deck. Tulio comes up after a while to help, since they are waiting to finish cleaning the presses. It's already been eleven hours of work today. Finally, we get enough netting off to move the tractor and I drive it down to the press deck. It's cleared enough to use it; we'll fix the rest tomorrow. Everyone chuckles about the problem, and later I tell Jamie not to be upset—everyone makes mistakes. We just fix it and go on. Then a little later, I tell him that we all had a good laugh, and he should, too.

"You all laughed?" he questioned.

"Yes of course! It's funny and now it's over."

He smiles.

Sept. 29: The stock market crashes, and I lose one-half of my retirement savings.

Harvest through October: Harvest is always non-stop work. We bottle to free up tank space, just in time. And then we race to harvest and press each day. I ask a neighbor, Fred, to help me bottle champagne. I am in the maniacal state, doing two things at once, going at a frantic pace. I get the new cases, take out the bottles, and push them onto the filler. My right arm hurts constantly from the motion. I ignore it and keep going, pulling the bottles off as fast as they fill, and replacing them. The filled bottles are backing up for the bottle capper. I shift Fred over to loading the bins and work the bottle capper until we are caught up; back and forth for four hours straight. At 1:30 p.m., I send Fred home and go to the press area for the Viognier pressing. This is the first time we got a full crop from our Viognier. It's taken five years to cut back and grow healthy new trunks on vertical shoot positioned training, eliminating the failed experiment with double cordons.

Rolland arrived in mid-September for stone wall finishing and repair work. He has a gate opener and is usually working by 6:30 a.m. each day. One day the gate is left open in the morning, and when I go out to check on Rolland, I see a neighbor talking to Rolland. The neighbors think nothing about walking onto my property. and talking to people I hire to work.

"Hi John, what do you want?"

"I was just talking with Rolland about some work."

"I pay Rolland to work. You'll have to leave."

He keeps talking to Rolland, and I interrupt.

"I mean it John; you'll have to leave."

"Then I'll talk to you," he says.

"No, I don't have time."

He glowers at me. "You don't have time?"

"No."

Then he says he just wants to talk about bird netting.

Again I say, "You'll have to leave."

In disbelief, he starts moving.

As I walk away, he turns direction and walks back to Rolland.

I stop and watch until he finally walks out the driveway, and the gate is closed.

* * *

Each day we start at 7:00 a.m. and work non-stop. I work the same all weekend to catch up, clean the tanks, and rack the new wine. My right arm is quite sore from bottling, and the motion required to clear the crushed grapes into the paddle area that feeds the must pump.

Every night, Stars hunts the vineyard. Every other night, I am out with her getting a raccoon or chasing deer.

Weariness is the theme now. No food in the fridge, eating hoagies and pizza and knowing this is not good, but having no time.

October 18: Finally we are at the annual champagne tasting event. We have the new champagne labels to introduce. For the past two months, we have travelled to Steve's studio in Lancaster every Tuesday evening to work on new champagne label designs. The J. Maki branding is now launched.

We changed the format for the event this year. We had to. Last year, a customer ran into a customer's new car, and tried to leave. We had to call the police. A lovely young women decided to take her clothes off and parade around, to the delight of many. Another group decided to start a bonfire behind our house. There were other highlights that told us we had to change, and so we did.

This year was RSVP only, and everyone paid a fee. Attendance was a third of the prior year, with 75% of the sales.

October 21: The red grapes are coming in. It's Merlot

today, and the tanks are ready. We crush till 6PM. I clean up the equipment and finish at 8:00 p.m. No food in the fridge: crackers and soup will have to do.

October 22: We bottle in the morning and rack the whites. Then, I take the tractor and cart into the vineyard to pick up lugs of Merlot. This will go to the freezer for this year's ice wine experiment.

October 23: The maddening pace continues. One more bottling to finish the whites. Then racking, cleaning tanks, and containers.

The economy has tanked; sales have stalled since the stock market crash in September.

I've been reflecting on my time here, after Fred's passing. I am only now seeing my mistakes in being too generous, assuming everyone was caring and helping when really, they were all helping themselves to everything they could. It's so hard to see clearly when you're in the middle of it.

October 25: It's Friday and I start at 7:00 a.m. cleaning barrels, filling barrels, cleaning tanks, pumping from one to another all morning. Tulio, Armando, and Clara are picking the Cabernet Sauvignon. At 4:00 p.m. Tulio has to leave and Armando thinks it's too much for Clara. He'll come back in the morning to help me. We start collecting our gorgeous Cabernet Sauvignon grapes and at 5:00

p.m. Armando says they will stay. We finish at 6:30 p.m.. I rehydrate the yeast and add the yeast nutrients and tannins to the containers and carefully cover it with a tarp. We're almost done harvest. The remaining Cab will be picked next week.

I feel pretty good about the harvest. We did better this year all around, and with more bird netting, we can tighten up varmint control on the upper corner next year.

October 29: I must disgorge champagne by myself. I need a finished inventory for the holiday season. I figured out how to eliminate the ice bath for the bottle chilling step. I moved all equipment to the coldest room and stacked ten cases of riddled champagne in the same room, so everything is at the same temperature before the next disgorging.

November 5: The Cab is crushed and added to the batch from last week. We got one and a half tons. We're finally getting the production back after the vine kill in 2004. I add more yeast and say goodnight.

November 6: We start to take down some bird nets. We start at the top of the vineyard. We pull the net up and over the vines to one side. Then we lay the net out on the ground between the vine rows, and finally roll the net up.

November 7: Jamie has contracted with Keystone Outdoor for a billboard in Philadelphia. He has experience with the same in one of his business endeavors with great success and is providing the capital. We've been working with Steve in Lancaster on the design over the last month. The salesperson is here today saying the final artwork is due today. I check the paperwork, and we have till Monday, so we send her away.

November 26: I've been having a lot of pain at night. At daybreak, I hold my head and brace for intense muscle spasms, as I slowly roll left, let my legs drop off the side and onto the floor. Then, I push myself up with my left arm till I'm in the sitting position. I compose myself for a moment, and slide down the bed to the footboard, then use that to help lift me to the standing position. Breathing slowly, I get the neurostim unit on as fast as possible. It's immediate help, but the rash is getting worse from the sticky pads. Soon, I won't be able to use it any more till my skin heals. I go out to open the gate at 7AM and walk slowly back to the house. I know I will feel better later, but for now I feel awful.

November 28: Today is the start of ice wine pressing. Jamie will have to go get the frozen grapes by himself; I can't bend enough to get into the car.

The workers are here today to help with champagne disgorging. I ignore my body's aches and pains and get the champagne equipment setup. We disgorge champagne all morning and get the ice wine pressing started. Jamie will have to carry the buckets of ice wine juice into the winery every two hours until midnight, when we shut the presses down.

December 3: Each day is the same. I feel awful till around noon, then a little better, then pain all night long. Jamie asks, "How are you feeling? I didn't know about this condition. Why not?"

"I did explain this before."

"No," he rebuts forcefully, "You didn't tell me you had this disability."

I struggle to explain it again, and then finally say, "I don't want to talk about it. I don't want to think about it. I have to live with it and I don't want to be reminded about it. So, I don't talk about it. I try to take care to avoid this type of occurrence."

"How bad is this?" Jamie asks.

"Not so bad. When it's really bad, I am in bed and can't move for three days. I take a steroid dose immediately, and on the third day I can usually stand, then it's a slow recovery. This time it's prolonged, but I can still function."

"You didn't tell me this," he says. "I was suckered into this situation, and now I have no choice."

I see his point but assure him that I will recover. Then

I explain what has happened since Fred was diagnosed, and how there wasn't enough time to do it all. So, slowly I gave up doing what I needed to do for myself. I really haven't been trying to mislead him. Truthfully, I thought I had gotten to the point of maintaining myself without a major incident. This last six weeks took me a bit by surprise, but still, it seems normal to me, although, I can see his point.

December 4: Jamie gets the ice wine grapes again, but I can help at the press deck all day. The holiday season is starting slow. Wineries follow retail trends. Typically, 50% of the year's sales are in the fourth quarter and 25% are in December.

December 5: We continue with punching down the reds: Cabernet Franc, Petite Verdot and Cabernet Sauvignon are now all inside fermenting in bins in the barrel room, which is about 55 degrees F. I disgorge champagne for the first time in a week and keep the workers busy. I'm the main support for two families.

December 14: It's Sunday and I'm starting to feel a little depressed. The economy is terrible, and sales have been very meager. At this time of year, it's depressing not having visitors/customers to the winery. We have the winery and entrance decorated for the holidays, and it feels like we're throwing a party, and nobody came.

Sunday is the only day of the week we have a little extra time in the morning, since the winery is open from 10:00 a.m. to 5:00 p.m. daily, and we are the staff. Jamie and I talk about what got us each here. I remember being five years old and taking off on my skates in the creek out back just a little farther, around the next bend, till I was five miles from home. A police officer saw me at an underpass as the day light was dimming and brought me home. "Why did you go so far?" Everyone asked. "To see what was around the bend."

In junior high, I remember being at home bored nearly out of my mind and thinking there must be more to life. In college, I had a Volkswagen Idiot Manual for my Karmen Ghia, and I worked through each chapter. When I got to Engine Noises, I was convinced mine had one, and I took the engine apart. Someone asked me, "Why did you do that?" "Because I wanted to see what was inside the engine."

In my thirties, I remember wanting to reach my potential.

December 16: We leave at 6:30 a.m. to pick up the frozen grapes. It's an hour and a half round trip, and a little more in wait time at Denver Cold Storage. By 11:00 a.m., we have the presses loaded and start adding air pressure slowly; pausing every few minutes until the press is holding three atmospheres pressure. The drops of ice wine juice start flowing then, and by 1:00 p.m. the first bucket is full. The fourth bucket is full before bed–that makes ten gallons. The next day we clean the presses and set up for the next press day.

Thursday, December 18: Today we repeat the work of Tuesday.

Sunday, December 21: Time is running out for holiday sales.

Monday, December 22: It's bitter cold; one degree F. with wind chill, and no customers.

Tuesday, December 23: There were good customers today.

Christmas: One of the four annual days off.

December 27: A few customers yesterday. Today is slow, but we're still hoping for a strong finish for the year.

December 28: A few visitors, but sparse sales. I wonder why more neighbors and "the community" don't stop by to purchase one bottle of wine for the holidays or at least a bottle of champagne. Why wouldn't they want to support a local business? It feels personal that so few people stop by.

2009

On New Year's Day, we take stock of our situation with the vineyard work, the winery sales, and develop a plan for the year. The winery sales have been trending down since 2006, when the economy showed signs of softening. We will do more on marketing, promotion, and special events.

Monday January 5: Today is the last day to get the frozen grapes. I'm up at 5:30 a.m. and anxious to leave. It seems Jamie is slow. In the car Jamie asked why I am in a hurry today. I realize it's the doctor's appointment at 8:45a.m. I always get nervous before a doctor's appointment.

My blood pressure was 160 over something. I didn't hear the second number. We discuss that, and I say I can monitor at home. He reminds me of a few things. "Yes," I say, anxious to leave as soon as possible.

January 16: It's four degrees at 7:00 a.m at the back of the house. I take Stars out for a walk and cut it short with bone chilling cold stinging my eyes.

Later, I check email and get this note from Deborah

Brenner, author of *Woman of the Vine*. She had been working on a festival for women winemakers featured in her book in Philadelphia. I was invited to the event last year at Flemings, and Deborah is planning to interview me for inclusion in her next book update.

The following is excerpted from her email:

Dear all Women of the Vine,

I am very sad to write this letter. I have been working so hard to resolve the issues that arose suddenly with the PLCB. To my dismay, I must cancel the event in Philadelphia and move it to NYC.

As you know, I have been in discussions with Tony Jones and the Director of Marketing, Jim Short since July 2008. Tony asked me to bring the first Annual "Women of the Vine" event to Philly when I met him at a tasting in Atlantic City. I told him I was kicking off a tour on the East Coast in honor of Women's History Month, which started 30 years ago by a task force of women in Sonoma!

I drove to Harrisburg, Pennsylvania, several times, had many conference calls with all, set up seminars with the Philadelphia Wine School (Keith Wallace), teamed up with Larry Hill, Regional Director of the National Coalition of Ovarian Cancer, as our local charity. We had support from UPenn and my alma mater, University of Delaware. We even had Philadelphia's Mayor Michael Nutter attending along with former Philly resident and now winery owner, Princess Ann Marie Borghese. A radio spot was booked on WMMR, and we were arranging winemaker dinners at top restaurants in town.

To my surprise, I received a call last Wednesday that the PLCB would not share with me the list of people they were selling tables to for the tasting. Tony told me that he would handle the vendors and I am not to be doing it, even though I was paying UPenn for renting the space and organizing the weekend. It appears that they were bringing in many distributors to pour and celebrities to sell their wine in addition to women that were sales staff and not women winery owners, growers, or winemakers.

All my best,

Deborah

I respond to Deborah that I'm not surprised, since I have much experience that the PLCB operates to protect its union machine, and the three-tiered alcohol distribution network established nationwide after prohibition.

January continues without a break. We press the ice wine grapes for the first two weeks. The compressor lines keep freezing as I try to coax the presses into working each day until 10PM. This year's experiment is with Bordeaux reds. The days are mind and spirit numbing.

Meanwhile, Rolland is knocking out the false cinder block structure in the basement. The builder said it was to support a future fireplace. When Rolland completes the demolition, he discovers that the entire structure was filled with trash from the construction. We also found this in a wall upstairs. It was revealed that the contractor was trying to hide the fact that the steel beam was a foot too short and didn't reach the exterior wall. So, they built this façade to hold the end of the steel beam; that's why this end of the house sags. We make a repair to prevent further damage.

February: I start pruning in the old Vidal, which is closer to the buildings; easier to get to for the start of this most grueling work. The electronic pruners are both working well.

It's also time to start disgorging champagne for the year, but I can't do it alone. I was able to keep up enough inventory for fourth quarter by myself, but it's not an option for this year's inventory. I decide to try teaching Rolland. He's almost finished repairing the mess in the basement.

Rolland and I work three days a week and slowly build a stockpile of finished champagne. During the first two weeks, every problem happens; bottles break, the cork doesn't eject from the machine causing the bottle to erupt, and the wire hood machine won't set the wire hoods correctly, using two for every finished bottle. Rolland learns about champagne as only one does if you make the product. It's dangerous, difficult, finicky, has a mind of its own, and acts alive. We never speak about it while doing it, and especially never say that things are going well.

I hurt my hand one weekend cutting off dead trunk sections in the vineyard with the big loppers. I have to use both hands and arms to force the cut and get pain up my arm. I tell Rolland we'll have to delay the disgorging schedule for two days.

The month goes by in a blur of weariness. The weekends are spent in the tasting room and more pruning. Worry continues about the economy, the mounting bills, and the lack of customers. I focus on just getting through the day, the week, the month.

March: Our beautiful new website is the highlight of the month! All the hard work over the past ten months has paid off. Finally, an esthetic match all around.

April: March goes out with an arctic blast, and April starts with winter weather again. Good, that will keep the bugs at bay, and the weeds slower to start.

April 26: The barn swallows arrive. We are delighted to see them. We can only wonder about their trip down to

Central America and back. Jamie gets the Kubota out, and mows till dark, scattering the insects in his path for the barn swallows.

I close the winery at 5:45 p.m., and as I walk to the house a hummingbird buzzes by. He's early, but the same day as last year. I rush inside to make food and get the feeders out. It always feels more complete when the barn swallows and the hummingbirds have come home.

April 29: It's all clear on the weather; the rain has stopped. The ground is fully saturated but draining quickly. We press on with the new plantings of Chardonnay. I keep pruning the young vines and start tying down the cordons in the old vineyard. The race is on. My hand has greatly improved, and I can use it again without the support bandage.

June 30–The weather is challenging throughout June with cool temperatures and light rains. The disease pressure is high, especially from Downey mildew. The vineyard looks clean so far. I shorten the spray interval and start leaf thinning earlier than usual. Jamie is deluged with a bumper grass crop and is overwhelmed with the job of mowing in the new vineyards. He has mastered close-to-the-vine mowing, reducing the trimming in the old vineyard by half.

July–The sun finally appears and dries out the vines. I

add sulfur to each spray mix to combat the extra disease and continue the shortened spray cycle.

We notice the barn swallows acting more territorial and chasing any birds away from the vineyard. We'll try to get all of the netting on in early August. We know how to spray the vines with the netting on, if needed.

September–After providing wine tastings to hundreds of people, I create a checklist for how to be prepared for a wine tasting, and add the page to our website:

Tasting Room Etiquette

The Tasting Room is a special place at the winery where you, the consumer, have a rare opportunity to taste wines where they are produced. The winery offers you a tasting of a selection of wines that showcase what is available for purchase, usually in exchange for a tasting fee or purchase. The winery personal are familiar with the wines and help provide information for the taster.

Tasting wine is experiential, which means that each time you taste there is an opportunity to learn–about the wines, the winery, and your own palate. Because it is a chance to experience something new, respect should be given for each taster.

And we always say, "Come to the tasting room ready to taste!" And that means a lot of things:

- *With a fresh palate, not after a big meal or drinking alcohol, chewing gum, or using any smoking products.*
- *No perfumes, hair spray, creams, or lotions with fragrance. Basically, you don't want any man-made fragrances as they may interfere with your ability to discern the subtle aromas and flavors of fine wine.*

October: Harvest is a struggle with intermittent rains and cool weather.

December 8: I get up at 5:00 a.m. to get ready for the frozen grape pickup at Denver Cold Storage. It's day four of ice wine pressing. I make coffee and open mail to discover a letter from Google Maps. Online I check out our listing, and discover they have the wrong address; it's a place 45 minutes away!

We've been trying to get to the bottom of this for months. Now, we know that our local phone company published a data file that is feeding the internet world and the GPS systems with a wrong address. For months, we've been hearing from customers about problems getting here.

Our local telephone company admits to an input error and offers us $25 for each of two years of incorrect listing. We estimate in the last month alone to have lost $1000 in revenue.

I work every day for three weeks to track down all

the incorrect listings and try to get them fixed. It's a nightmare.

December 29: The year end stress is mounting; waiting to see if any of our "friends" show up to buy a bottle of champagne. We offered a 15% discount in the email this week, but no customers yet.

New Year's Eve: We beat last year's sales by 11%, impressive in this recession year, but it doesn't feel good enough.

I dream about drowning and driving my car off Grove Road onto the driveway that becomes water, and I have to swim down the driveway.

2010

January 2: Saturday. This is Jamie's 63rd birthday, but he looks years younger. We talk over coffee and decide on a tour event approach for this year. It seems that everyone tries to make the tour and tasting a private party for themselves. It's no fun looking into the blank eyes of so many who have no knowledge or quest for it. They just want to have a good time. Who could think that champagne is actually fine and elegant?

We have decided to open at 9:30 in case of weekend travelers. The first couple arrives at around 11:15, from New York. They heard about the champagne and ice wine and want to taste those. The man says to Jamie, "She wants to sit down." Jamie says, "I'm sorry, we don't have seating." The woman storms out the door leaving the door open in 20-degree weather with the heat blasting out. The man follows her, leaving the door open also.

January 4: We talk about how to get distribution for the winery. In the past, I found outside work to supplement.

Now, it's "do or die" for the winery. It's got to pay the bills.

I go to our general practitioner for my finger; still feels like there's a sliver of glass in it from the cut two months ago. There wasn't time to go previously with harvest and the extra ice wine pressing work.

Dr. Donaldson is a pro—he had to dig around in my foot several years ago after a glass splinter. He examines, then anesthetizes the finger, and starts working on the red area. Blood gushes out, and finally he has to tourniquet the finger. After about 15 minutes, he closes the area up with two stitches, and gives me antibiotics for three days. I leave with a big bandage on the finger.

January 7: Another tourist sign is down on Route 23. This is the third one since we put them up. It's rush hour when I see it and decide to come back tomorrow to pick it up. The sign installer says that someone is cutting them off.

January 8: We had more than an inch of snow overnight and Rolland is here at 7:00 a.m. sweeping the driveway. After he finishes, we disgorge Blanc de Blancs.

January 10: Sunday—Stars is out early, and I see her run uphill from the new vineyard. She is chasing a red fox. The fox runs to the stone row opposite the compost pile; probably wants to get to the grape remains. Stars sits nearby and stares, the fox returns the stare, then breaks the look, and meanders back and forth over the stone row feigning indifference, until Stars breaks off.

I take care of a young couple who have seen the signs for the winery and decided to visit today. He likes sweet wine; she likes dry reds. Each time before he gets to taste, the young women says, "You won't like this," or "You;ll like this." I want to say, "Please, let him have his own experience."

Sunday, January 17: I'm cold and wet all day long disgorging champagne and pressing ice wine. We got an order from the PLCB. Jamie encouraged me to call and we're pleasantly surprised that they want to try the Blanc de Blancs and ice wine in 28 stores.

January 18: Martin Luther King Day. We've had decent traffic at the winery over the weekend. Jamie has been washing the '04 Blanc de Blancs bottles and labeling each day for the PLCB order. I have to figure out how to create the bar codes.

January 23: It's time to start pruning. The electric shears are charged, and I take the sharpening stone to the vineyard. I work on the first row of Gewurztraminer. After two vines, I stop and look around, remembering the long-haul pruning is. My body already hurts everywhere. The thought of pruning 16,000 vines is overwhelming. I finish the first row in about 90 minutes; that's 50 vines. And this doesn't include the work to remove the old cordon and trunk sections and the final tie-down of the new cordons. That's at least another 90 minutes. That's three hours for 50 vines. So, for the 16000 vines that's 960 hours. And two people have 4 months—from January to May 1st—to complete the job regardless of weather.

January 26: I go out to the vineyard with Armando and show him where to cut out the old trunk sections. I'm always trying to renew the vine, little by little every year. I continue pruning, and get two more rows done in the morning, and one more after we close the winery. The light is just enough to work till 5:30 p.m.

January 27: I travel to the used restaurant supply house north of Reading, and get a surprise on the way; our champagne billboard has been moved to Route 62 and it looks great!

February 5: We're expecting an "historic blizzard" to start around 3:00 p.m. If the storm materializes, we may not be able to get into the vineyard for a week.

February 6: At 6:00 a.m., we have over two feet and counting. At 11:00 a.m., we go out to survey the damage. I head out to free the gate area, and Jamie tackles the front walks. There is drifting to four feet. We work all day, and at 2:30 p.m. Rolland arrives to help. I still haven't got the gate area cleared to the road. He works until 4:30 p.m. and then calls it quits. Jamie and I drop onto the floor at 6:00 p.m.. It'll be days before we get the cars dug out.

Later that night, Jamie wants me to rank in order of importance five things—all of them important: Janet, Jamie, wine sales, the vineyard, and Stars.

I rank them in that order. His ranking is Janet, wine sales, Jamie, vineyard, and Stars.

February 9: Armando continues to dig out the parking area. The 20 inches of snow on the tasting room roof now has an ice layer on the bottom. They're projecting another 8 to 20 inches over the next 24 hours. Two major snowstorms within a week have never happened before.

February 11: Finally, the blizzard is over. The plow comes down our road with a first pass and pushes the snow into our cleared driveway. Now, I must shovel that out again. As I work on that, I hear the plow coming back, and stand out in the road to block the plow truck. I'm hopping mad and yell at the driver. After the first storm he knocked my mailbox down. I say, "You have control of the plow blade. You can push it past my driveway. I just spent four hours clearing it.". "Sorry," he says. I decide to clear off the top of the stone walls on the roadside, so they are visible until the snow melts, hoping to avoid more problems.

February 12: We work all morning on a pathway from

the driveway into the winery. At noon, Jamie takes the first visitor. It's a township employee that comes to say we will be fined for putting snow in the road after the plow had cleared it. Jamie asked him, "Where should I put the snow?' and the employee says, "I have no idea—just don't put it in the road".

February 14: Sunday, Valentine's Day. We start the day slowly. We're all sleeping in the living room this past week since we got snowed in. Stars and I are on the floor, with Stars on top of the quilt and Jamie on the couch. We all fall asleep about 8:00 p.m. I am restless most nights with body aches and must move from side to side to back all night long. I try to be *native* and practice moving silently. Jamie always lies motionless on his back since his cancer surgeries in 2004.

February 21: We have a few customers. That night we declare that 2010 is the Year of......

Curtains

Relaxation

High Performance Bodies

A New Van

March: When I began the vineyard and winery, I hoped to create local wine that the local community would be proud of and want to enjoy. So, I supported every request for a donation for eighteen years. Only once, in twenty years, did anyone ever buy a bottle of wine when they came for the donation.

So, I began to realize that it must be that they just don't understand how it is here; how much work and years of effort it takes to create a world class wine, or they wouldn't ask for donations. They would ask, "How can we help you? We want you to stay in our community."

I did countless events and tours of the winery and vineyard and opened the historic log cabin for house tours and for Chester County Days, so "the community" would come here and learn how it is. I showed those interested what we do, talking about how only three of us farm 16 acres of grapes, about 16,000 vines individually tended; working every day of every year, to farm sustainably; then we make the wine here, and retail it here, and how everything is done by hand, and how we work other jobs to support ourselves and the winery so we can bring good local wine to the community; how we work tirelessly to make the property attractive so "the community" can be proud of it.

But the donation requests keep coming. It must be that they don't understand.

Then, I remember that the Chester County Community organization was coming here for a meeting last fall, and how I prepared for the visit, and looked forward to talking about what we do here. That morning the meeting was cancelled. I remember how I was disappointed but was told "It will be rescheduled." But it wasn't. I guess they just don't understand.

April 19: I'm tapped out from pushing all week to get the pruning done. The electric pruners feel heavy, and I'm getting bad headaches all night, a residual problem from the severe neck injury decades ago. Jamie takes the load in retail again, giving me a rest. Every night, I sleep on the living room floor now, and Stars keeps me company. She stirs when I sigh with pain. I can't lay on my side; getting up is a monumental ordeal. I breathe fast and shallow holding my side, a residual from the cracked rib. Late afternoons, I can't wait to lay down and go to sleep. After an hour laying down, I lie awake most of the night, and can't wait to get up.

April 20: I'm feeling better and work in retail all afternoon doing little jobs. Jamie gets to ride his new small Kubota mow tractor. Another 0% down arrangement. He loves it. He's been working hard for three years, supporting me, and getting nothing in return. The hope of our future together where we don't have to slave all day, every day, keeps us moving forward.

April 21: **Faithless is he who quits when the road darkens.**

This is my life. Having faith that I am on a road to somewhere. Faith isn't only a religious entity; it is a belief in oneself.

We will start the replanting this week; planting new vines to replace vines that have died.

April 24: The days are busy pruning and tying down the new cordons and picking up the clippings to make the burn piles. We learned many years ago that removing and burning them greatly reduces the disease pressure in the vineyard.

April 26: Last weekend, Jamie had a returning customer that said he knew our U.S. Senator and offered to talk with him about our Vinalies award. He knew the Smithsonian Institution had the two California wines that were the first American wines to win in Paris. Since we were the first non-French winery to win gold for champagne, that accomplishment was something the Smithsonian would be interested in. Today, we sent an application to the Smithsonian, and scheduled a meeting with our Senator.

April 27: The sprayer returns, and the tractor hydraulic arms go up and not down. I made a fatal error, something I have done before, and kick myself for having to learn this again and again. Never assume that the company or representative that services your equipment knows anything about operating it, or getting it attached to or from your equipment. It's the same thing over and over; the only way is to learn how to do it myself.

We decide to get another Odyssey for Jamie who gave up his car 2½ years ago. He will have his own car then and can take off any time he wants. I called the salesman last week and he said to give them two more days to locate one. There are no customers today. Jamie mows till dark.

April 28: I finish the Merlot pruning today and start pruning the Syrah. No customers again. Jamie mows till dark. The good news is they found him a car!

April 30: Customers come from the turnpike billboard and buy a lot of ice wine. Maybe this is our salvation—ice wine. Ice wine is sweet, and the general population likes sweet beverages. Jamie is upbeat and likes his new car. We start planting the Petite Verdot today.

May 1: Today brings steady traffic in the winery, and better customers; not just the couples looking for an activity. Jamie and I work together all day, making sure we each get a break from the customers to do some outside work. He continues mowing and I continue finishing the pruning.

May 3: I call the propane company and ask when the new tank will be here. I was told eight months ago. The woman bends over backwards to be accommodating. Since the economy has been very poor, customer service is back in vogue. "Monday," she says, "We will be there."

May 8: Saturday sales are modest the day before Mother's Day. Later in the day, Jamie and I talk about how I wear myself out giving everyone an "all experience." I try to educate and give them a personal tasting experience. But most just want some alcohol, a break in their day, and maybe they'll buy one bottle. We talk about how to do this better. Jamie gets the customer talking about what interests them, and then targets his presentation specifically to that. My approach was always to be informative. I understand now that his approach is more successful.

May 9: Sunday, Mother's Day. Traffic is average; we're a little disappointed. With two billboards, we should be seeing more, but the month's numbers are good overall. Every day is important, and we have to do our best all of the time. Yesterday, I drove out to East Earl to buy annual flower plants for our pots. I got all the flats into the old van, with Stars, my dear companion. I was worn out driving home, and aware that the hitch in my breathing from the cracked rib was not going away. At home, I have to open the winery after unloading the plants. Jamie is finishing the trimming on the roadside at the entrance. He's doing an amazing job as vineyard floor manager!

June 16: Today is the 19th and Chestnut Street PLCB store visit. Our champagne isn't on a shelf. I ask the manger to please look in the back room. There are piles of cases, and we look through them all. I had to do the same thing in the Media store. The champagne has been in the PLCB stores since March. We look for an hour and can't find it. I say to the manager, "You have to move that one stack to see if it's behind it." She tells me that's a lot of cases and they'll look for it when they have time. I'm thinking I'll have to stay no matter how long to find it; otherwise, it will never be sold. I haven't had any food or water since 9:00 a.m. It's now 3:00 p.m. and I'm fading fast. I'm at my wits end and say emphatically, "If we don't sell this, we can't pay our bills. We must find it. I can help move the stack." She agrees and gets three others to start moving the stack. Finally, we find the one case, and get it on the champagne shelf. I say I'm sorry for making such a fuss. She says that's okay, and I leave thanking everyone on the way out. I remember the difference in what my visit must have looked like, compared to the three business suited distributor reps in their thirties, when I walked in. They were walking around checking their products already on the shelf and smiling to everyone.

Back in my car, I start driving to the next store. I'm almost 60 and can't believe I'm doing this.

June 17: I finally finish all the trips to visit the 28 PLCB Wine and Spirit shops that have our Blanc de Blancs and Vidal Ice Wine. Driving is so uncomfortable for me; after two days, my body hurts for a week. It's taken six weeks of concentrated effort to complete the visits just to get our products on the store shelves. Without this effort our products would sit in the storage room and never get on the shelf.

June 20—Jamie's two sons arrive for dinner. After dinner, Jamie tells a story he has never told anyone before, and he wants his sons to know.

He was nine years old and set up a Kool-Aid stand in Lancaster, two blocks from his house at a major intersection. He was alone. He determined that sales would be higher if he served multiple flavors. Today, a man in a pickup truck stopped and asked to try the flavors so he could pick which one. Jamie had little Dixie cups for tasting and gave the man four tastes. The man then picked one of the flavors and Jamie gave him a glass, which the man drank, said thank you, and walked away without paying for the drink! Jamie was stunned and mad, and already had a good pitching arm. He looked around for something to throw. He noticed the cinders on the side of the road from winter, scooped some up and threw it at the man's pickup as the man drove past him. He still remembers the sound of the cinders scattering noise as they splayed into the side of the truck. After a moment the pickup stops, and the man looks at Jamie, who grabs another handful of cinders and makes a start toward the truck. Quickly the truck speeds away. Jamie calls it a defining moment in his life.

June 28: I wake up at 2:00 a.m. thinking about today's bottling, the '07 VSP, our best ever. It's intoxicating. Have I done everything possible to insure it gets into the bottle in perfect condition? The shepherd's job doesn't end till it's in the bottle.

July 2: I leave for the Y after lunch for some unpressured time and notice a cigarette box littered on the Belgian block at the driveway entrance. Marlboro light; same as all the other times. This could be from another jealous neighbor or someone harboring ill will. If you are in Pennsylvania, the more you seemingly possess, the louder the community unrest.

July 8–We receive this email from the Smithsonian:

From: Johnson, Paula
Sent: Thursday, July 08, 2010, 4:27 PM
Cc: inquiry
Subject: re: an object for consideration

Dear James and Janet Maki,
   Thank you for your e-mail of 26 April to the Smithsonian's Public Inquiry Mail Service regarding a possible donation of your award-winning 1997 Blanc de

Blancs Champagne to the Smithsonian's National Museum of American History. We appreciate your interest in the Smithsonian collections and wish to convey our sincere congratulations on being awarded the esteemed gold medal by the Vinalies Internationales in Paris.

While we do collect documents and objects relating to the history and culture of winemaking in the United States, we generally do not maintain collections of organic material. The only exceptions involving wine are the vintages that placed first in the 1976 Paris Tasting–the 1973 Chateau Montelena Chardonnay and the 1973 Stag's Leap Wine Cellars' Cabernet Sauvignon. The long-range impact of this event, and its unexpected outcome, are widely acknowledged as a turning point for wine and winemaking in America. For this reason, we made an exception to our general collecting guidelines.

Thank you again for your kind offer, and we wish you continued success!

Best regards,

Paula J. Johnson
Curator

Division of Work & Industry / MRC 629
National Museum of American History
Smithsonian Institution
Washington, DC 20560

July 28: Last day as a 59-year-old.

Problems have been brewing in the vineyard. Perfect weather turned to daily showers for the last two weeks, and we're into an overgrowth situation.

September: By the end of September all the varieties are picked, the whites processed, the ice wine grapes are in the freezer, and the reds in fermentation. The red berries are so small, and the liquid so small, that there is barely enough juice to cover the solids. We end up with a single barrel of each of the red varieties, but the pressed wine is already better than any previous harvest.

Traffic to the winery continues steady during the week. Harvest brings in the tire kickers and the would-be's that have visited "Sonoma" and know it all; especially, "that you can't make good wine in Pennsylvania." It's incredible to me that people don't acknowledge that there are always exceptions to every common perception.

December: We wonder what the economy will do this month and how we will fare. The first two weeks are bleak. Then it picks up starting the 14th and goes strong through Christmas and New Year's Eve. We're open every day from 10:00 a.m. to 5:30 p.m. We end up 20% YTY; incredible with this economy, but the PLCB order was responsible for most of that 20% YTY increase. Hopefully these orders will continue.

Progress is visible all around for the first time in 20 years. We have new gardens, a new house porch, new cars, returning customers, new activities, new retail help, and new hopes.

2011

January hosts extreme cold and snow every five to seven days, usually about three inches. In order to start pruning this year is I have to tamp down a path to the top of the vineyard. The sun on this south facing slope helps to melt the snow at the top of the vineyard first. This makes it a little easier to work in the rows. One day Jamie decides to film my daily climb to the top with our new video camera. We have ideas to make videos for our website of the various vineyard tasks. At the end of the month, we start disgorging champagne. It goes agonizingly slow. We work all morning and finish only six cases, a third of the usual production. We're working on the last of the 2002 Blanc de Blancs, aged seven years on the lees. The lees add complexity, so this is precious stuff. Our Blanc de Blancs (white of the whites) is 100% Chardonnay. I have come to believe that it is the most difficult of the traditional champagnes to finish. In my visits to France, and discussions with French producers, they have shared with me that there is no requirement that Blanc de Blancs be 100% Chardonnay; therefore, many producers add some Pinot Noir to the mix, which "calms" down the effervescence. I prefer the maximum liveliness of 100% Chardonnay and know it will be more difficult to finish. Any Pinot Noir will mask the delicate, creamy, yeasty, whip-cream like aroma and flavors of our Blanc de Blancs. Pinot Noir *must* for champagne is pressed off the skins and seeds and fermented like a white wine. The *must* has a slight pink color and the perfume and aromas of strawberries, and more tannin than the Chardonnay. I now believe it is the increased tannin which calms down the effervescence.

We're working in a 32-degree room, and everything is wet. About 10:30, I say my toes and fingers are starting to feel frozen. "Yeah," Rolland says, "and I'm wearing steel toed boots." "Why don't you wear the muck boots I bought you for this purpose?" "Oh, are they better?" "You'll have to try yourself and see."

The second day is no better. I had cleaned the corker jaws, but they weren't very dirty. This vintage is super active. After three cases, I say I can't take it, and we switch to another vintage. It goes a little better.

I alternate days in the vineyard with days disgorging champagne. By the end of February, we have a small cache of finished champagne, and the snow has melted in the vineyard. Vineyard pruning takes priority most days. Rolin can only help out on Sunday mornings with the champagne disgorging work. By mid-April the work room is warming up with the outside temperature. This advantage of a cold room ends.

I now turn my attention to the vineyard replenishment project for this year. My approach has always been to nurture new trunks from the graft union area. Over time some vines die. Equipment injury or winter damage are the usual suspects. I have heard that vines only last 15 years for hybrids and 30 years for all others. When I hear someone say this I ask "How is it that the most famous European vineyards are 100 years old?

I remember discussing this with a 75-year-old vineyard owner during my last trip to Sonoma. He affirmed my belief that as long as you care for a vineyard, the vineyard will go on. I believe that people say a vineyard lives for 30 years because that is the length of time one person usually works at it. If you start a vineyard in our twenties, you end in your fifties after thirty years of hard work.

This year I will start the Chardonnay replenishment program—50 replacement vines for those that have died over the last 15 years. We will have to dig each vine in by hand and protect it for a few years as a new trunk emerges. It takes longer for these replants to bear fruit as the mature vines next to them have established roots and block them from 'sun' space. It will be five years before we see the first fruit.

I have been monitoring the PLCB store sales of our Blanc de Blancs and Vidal ice wine. Each store has one case (12 bottles) of each. The PLCB listed us in their 'luxury' wines, which we found out meant that a reorder would only come after the individual stores requested a reorder. The PLCB manager for Pennsylvania wines suggested we do in-store tastings of our products. The distributor reps had regular time spots but there were some open spots that we could get. I began to schedule a state store tasting every two weeks. The holiday reorder decision would be made in September, and we had to sell enough product by then. Jamie and I take turns driving to a state store on Saturday afternoon starting in May while the other person works in the Tasting Room at the winery. We must take a table, glasses and chilled product to each store. We manage to get to 14 stores before harvest starts. We usually sell between 1 and 5 bottles at each store, but we don't get a reorder that fall.

We stop into PLCB stores close to home whenever we can to check the inventory. The Exton store manager was very friendly and has our wines on the top display shelf. He encourages us to come back in December for another in-store tasting, and says he personally talks to many customers about our wines and our nearby location for visiting. A few other store managers have also been very supportive, and we decide to schedule another in-store tasting at Exton, Gateway, and Wayne in December.

Harvest goes quickly with good weather and low yield again. The vines were stressed last year with low growth, and it shows in yield this year. More grapes are in the freezer as the Ice Wine is gaining in sales since the new billboard.

2012

January 4: I checked my electric pruning shears last week and sent one of the shears off for repair. I put on all my three clothing layers, plus the battery pack. I must climb to the top of the vineyard, to begin the most daunting task we have—pruning the 16,000 vines. I make the decision cuts on the old cordon and clear the trunks. Today, I completed two rows; starting is the hardest.

January 6: After lunch, I go out to start pruning in an area close to the house, so I can watch for customers. I force myself to do the rows in order. The Vidal is the hardest to prune, and I want to avoid it. I have to finesse the vine, cut it back hard each year, and still have some shoot capacity to grow grapes. It's a miracle every year that they consistently produce. The prevailing wisdom is that hybrid grapes varieties only live about 15 years. I was shocked to learn this after growing them for 16 years. Our Vidal (French American Hybrid) seemed to get stronger when I converted the vineyard to permanent under the vine row vegetation, eliminating herbicide banding, which is the standard in vineyards. Last summer, I attended a Penn State seminar in Monroeville that showed the result of ground cover to reduce the *nemotodes* (plant parasites) that damage the grapevine roots. I believe that is why our Vidal lives on and produces.

January 25: I've been pruning daily except for three snow days and one rain day. I'm nearing the month's end short of my goal of rows completed, but better than ever for January. It's incremental improvement.

Today, it's 30 degrees in the tank room. We spent a day last week getting ready to disgorge champagne. It's so much work; it's so cold in the room and I am all wet during the work. I'll wait to start in February. All this work, outside and inside in the cold and wet, is damaging to your health.

February: Jamie and I decide on a monthly event schedule to bring more customers to the winery. In addition to the spring pruning workshop and the Champagne making tour we decide to offer a special event each month this year:

- Wine Tasting 101 For this event I prepare samples of aromas and flavors common in wine such as apple, pear, plum, grass, tannin, and charcoal. The second part is tasting samples of wines with different sugar levels (0.25%, 0.5%, 0.75%, 1.0% and 1.5% samples). Each person learns their own sugar threshold. The human threshold is 0.5% Residual Sugar (R.S.). This means that this is the minimum amount of sugar needed for a human to detect, and most people can't detect sweetness until the sugar level is 0.75%. The sugar level is the result of adding sugar to the wine. Technically, dry wine means the fruit sugars have been used in the fermentation, and none is added back in.

- Champagne Dosage Tasting. The dosage is the finishing liquor added before the champagne is corked. Every champagne house (winery) uses their own formula. We found that our Chardonnay was the best base liquid to use for the Blanc de Blancs dosage. For this tasting event we made samples from different vintages of Chardonnay and different sugar levels in the Brut classification of 0.5% to 1.5% R.S.

- Vineyard Tours The early season tour would showcase the new buds and shoots in the different grape varieties. The pre-harvest tour would showcase the full canopy and grape clusters prior to netting.

- Harvest Workshop on Saturdays during harvest we demonstrate the various harvest activities.

- Tour and Taste the Grapes and corresponding wine in the vineyard. This event would take the customers into the vineyard to taste the grapes on the vine first and then the finished wine. The tasting included: Chardonnay, Blanc de Blancs, Gewurztraminer, Pinot Noir, Merlot and Petite Verdot.

The winery now has two Pennsylvania Turnpike billboards, two Turnpike exit signs, and eleven tourist signs within five miles. Also, twenty-eight Chester County tourism signs around the county. The Turnpike exit signs require that we are open seven days per week.

April: Pruning is slow with only one worker to help. The new tours are getting a lot of interest.

May: First time ever, we had a killing frost in early May. Most of the new buds were killed. This will decrease the crop; each bud produces two clusters of grapes.

June: We invest in a new roof for the old milking shed.

August: After all the advertising, the Ice Wines are the biggest sale item. We have a good inventory of table wines from the plentiful years of 2006-2008 and decide to use most of this year's crop for Ice Wines.

October: The crop is small, and all varieties are picked earlier than usual. I decide to give the oldest Cabernet Sauvignon extra time on the vine (called *hang time*). It's in good shape and reaches 24 Brix. We've never seen that high in Brix before. Brix is the measurement of sugar in a solution. The Brix measurement is used to estimate the alcohol level of the finished wine by multiplying Brix times 0.59. So, this vintage of Cabernet Sauvignon will be a little over 14% alcohol, which is unheard of in Eastern U.S. wine. California is known to produce 14% alcohol wines on a regular basis. Higher alcohol produces a *hot* flavor in the wines.

December: The holiday traffic is slow. Year over year revenue is down 20%.

2013

January: We decide to continue all advertising and the event/month schedule for this year. We'll take a break from the PLCB in-store tastings until springtime.

February: I work alone disgorging champagne—12 cases per week. The cold room is too much for me working alone—it's too long to be wet in 30 degrees F. I move the equipment into the barrel room. Without heat it stays around 40 degrees. I set up 12 cases a week in advance so they will be the same temperature as the air temperature. I discovered that was a key factor in keeping the effervescence calm enough to finish corking. If I keep at it, I'll have over 100 cases finished by early April when all time goes to finishing pruning.

In March, we get a call from the Jimmy Rollins Foundation. Jimmy Rollins was the star short stop for the Philadelphia Phillies. The representative says they are aware of our champagne award and would like us to be the champagne and wine sponsor for the Foundation event in Philadelphia that June. The Phillies will be at the event and local media. We will get publicity, two tickets to the event and to the sponsor reception with Jimmy Rollins, an autographed ball, and other memorabilia. Jamie and I both enjoy baseball and are Phillies fans. We decide to say "yes."

I finish pruning the Cabernet Sauvignon in early June. It's been wet, cool, and cloudy all month. The disease pressure in the vineyard isn't evident yet, but I'm preparing for an all-out assault. I decide on a 7-day spray interval instead of a 10-day interval, with extra Botrytis chemicals and Sulphur. This was a lot more work, while keeping up with the vine canopy growth. Again, our cane pruning pays off with modest extra leaf growth. Had we stayed with the bud/spur approach we would have had three times the leaf growth. With the wet weather the leaves were covered with Downey Mildew. We were constantly pulling off diseased leaves and trying to open the canopy for wind and sun.

The Jimmy Rollins Foundation representative calls and says they can't pick up the champagne and wine and that we'll have to deliver tomorrow afternoon—that's Friday afternoon June 14, in Philadelphia, before the Saturday event. Someone will be there to unload. I'll have to go as Jamie's hip is starting to be a problem.

On Friday, I load 6 cases each of Champagne, Chardonnay,

VSP Bordeaux blend, and the Vidal ice wine into our old Dodge caravan. I know Philadelphia well and find the dinner theatre near Front Street easily. I park in back near the loading dock. No one is there to help me unload. I find a hand truck inside and start unloading one case at a time. I load three cases onto the hand truck and take them inside and unload. I pray that my back muscles don't go into a spasm. I take short rests and try to stretch. After an hour I am done and start the long road home in rush hour.

On Saturday afternoon we get out of our farmers' cloths and dress for the Jimmy Rollins Foundation event. We arrive early for the sponsor reception which starts at 4:30 p.m. At 5:00 p.m. another sponsor arrives, and no foundation representative is there yet. At 6:00 p.m. Jimmy Rollins and his beautiful wife arrive through the back entrance accompanied by several media and TV personalities and cameras. The Phillies team members and their guests all gather around. I am surprised that most players are short except for the pitchers—who are all tall. After twenty minutes of questions and answers everyone disappears. We were expecting to meet Jimmy Rollins and ask the event representative. We're told that will happen after the presentation. The highlight of the evening was meeting Cole Hamels and Kyle Kendrick. They were interested to learn about our vineyard and winery. They said they don't often get to meet the people who make the products from their sponsors.

August: Cool and wet has been the overall weather pattern all growing season. I've been adding new chemicals to combat the risk of rot in the vineyard. The reds are slowly changing color- several weeks later than usual. The berries are in what I call "a stall." Too cool to develop color and flavor, and too wet to dry out. There's still time to ripen the grapes if the sun comes out, the temperatures climb above 80F every day, and there is no more rain. The dew is especially heavy in the morning causing the leaves to be wet all morning.

September: All white varieties this year are at risk with botrytis and other rots. Botrytis is romantically called *noble rot* and thought to impart a "richness or honey like flavor" to the wine. My experience is that it's impossible to distinguish between the types of rot, e.g., there's Botrytis, black rot, sour rot, bunch rot, fruit rot, and bitter rot. Most rots just make the wine stinky. Because of that we always pick out any rot or diseased grapes in the vineyard by hand, so they don't get into the crushed grapes.

We start to pick the Chardonnay and it takes all day to pick one row of "clean" grapes. The rot continues to affect more grape clusters every day. By weekend, we have only a small amount and take everything to the freezer for ice wine.

The following week we start to pick the Vidal. After one day we have one lug (40 lbs.) picked. Normally we would have three rows picked for a total of 40 lugs. After three days, we abandon this variety, due to the excessive rot.

October: The red grapes are slow to ripen. The skins are tougher than the white grapes and withstand the moisture pressure better. We decide to pick whatever we can at month end and make a red ice wine blend. Except for the Cabernet Sauvignon which is still holding its own.

In November we discover the deer are getting into the nets around the Cabernet Sauvignon and feasting all night long. We can't protect the grapes anymore and must harvest. The crop is small, but the color is intense, and the aroma is of ripe black cherries.

December: Customer traffic has been slow this fall. Revenues are down 20% for the year.

2014

January: We decided to continue all advertising, and to offer a champagne making tour at noon on the weekends through April, and champagne by the glass for $10. I'll be disgorging champagne in the cold months and will use the equipment already set up for the tour.

February: Our finished wine inventory from 2007 and 2008 is still strong. In 2007 we were blessed with a plentiful crop of outstanding quality. The more recent small vintages made into ice wines are more economical: the output is 50% less, and means less costs for bottles, closures, labels, and fermentation supplies.

Our first Cabernet Sauvignon ice wine is almost finished fermenting. Each time I lift the container top, I am greeted with a profusion of strawberry aromas, and it fills the room. I've never experienced this before. I decide to let the champagne tour customers experience this–and they can't stop talking about it. Many customers come back that Spring wanting to know when this first of a kind Cabernet Sauvignon ice wine will be available.

March: We contact our State Senator, whom we've known for years, for help getting a meeting with the new PLCB Chairman. The new Chairman doesn't go to his Harrisburg office, but instead works from his law office in West Chester. Our Senator knows a lobbyist who knows him and will get us in touch with him. We're still hopeful to get another state store order.

In late March, we get a visit from the lobbyist, and he is very enthusiastic about our products. He agrees to work on a meeting with the PLCB Chairman. The lobbyist's assistant contacts us the next week to ask if we are interested in donating to one of their charities. We are happy to oblige.

The lobbyist's assistant contacts us to discuss the meeting with the PLCB Chairman and suggests we host a lunch meeting at the winery. Then, he asks where to send the contract for their services. They require a $3000 per month retainer. We agree to try 90 days, and commit to the ongoing retainer, if we get a PLCB order.

The lunch for our Senator and the PLCB Chairman was scheduled for mid-May. I prepared a wine and food pairing lunch with our wines:

Field greens with champagne vinaigrette

Braised turkey breast with Gewurztraminer reduction
Homemade profiteroles and ice cream
Our Vinalies awarded 2003 Vidal Ice Wine

We had samples of three wines for state store consideration. During the lunch the Chairman asked if we were interested in buying a state store. We were in total surprise as this was not a publicly known idea. He asked if we could afford $200,000. We knew the going rate for a liquor store in NJ was about $400,000. We expressed surprise and were told that we would have a 'silent' partner in the deal. We replied that we wouldn't be able to do that.

The lobbyist said he would talk to the Chairman later in the week about our wines for state store consideration. Three weeks later we receive a letter from the PLCB Chairman thanking us for lunch and wishing us the best; there will be no PLCB order.

August: We continue to prepare the vineyard for harvest and decide to target all the red wines for ice wines. The Cabernet Sauvignon ice wine experiment was bottled; only 12 cases. Every time we open the bottle for tasting, the strawberry aroma fills the air. It is intoxicating. This first of its kind sells out in three weeks.

September: The white grapes are harvested by month end and have amazing fruit flavors. The ripening this year occurs at the lower end of Brix (sugar content), which means more delicate fruit flavors such as apples, pears, peaches, and melon. We're still getting small crops but after last year are happy to get a quality crop.

At month end we decide to offer a sale that we think can't be resisted, buy one bottle and get two free bottles. We sell more in one weekend than we had hoped to sell to the PLCB, and the sale brought back many old customers.

This fall we have more returning customers bringing new customers. The ice wines are taking center stage with the new customers. We add volume discounts: 15 % discount for a three-bottle purchase and 25% discount for a six-bottle purchase.

In December we offer our existing customers a special Private Customer sale at 50% discounts for mixed cases. We spend the weekdays labeling inventory. Existing customers pre-order 150 cases for pickup by December 15.

During the Private Customer Sale weekend, a customer's wife arrives to pick up her husband's order. She has brought her mother and wants to taste all the 20 wines on the sale. We're very busy but try to indulge all our customers. After an hour the customer's wife wants to add another case to their order. Her credit card is denied and after three attempts she asks to use the phone. As she stands in front of me, she calls her credit card company and starts yelling at them to approve this purchase. Her mother tells me that she always gets her way. Eventually the credit card company will not approve the purchase and the customer asks her mother to use her credit card.

We're amazed that on one weekend we sold 2300 bottles of our wines! We get cards from our customers thanking us for helping to make their holidays special. We reverse the YTY down sales trend and end up 18% YTY.

2015

January: Our new, first ever, red ice wines from Pinot Noir, Cabernet Franc and Merlot in addition to the second vintage from Cabernet Sauvignon are fascinating, with each red grape variety expressing their dominant fruit aromas. The aroma of the Cabernet Sauvignon Ice Wine is strawberries, the Pinot Noir is dark red cherries, the Cabernet Franc is blackberries, and the Merlot is red plums. We decide to provide a sample of these upcoming new ice wines for the customers who are tasting wines.

The customer whose credit card was declined has emailed several times that some of the wines were bad. She asks for replacements. We say "Yes, of course, just bring the bottles back and you can pick what you'd like in exchange." After some back and forth she says all three cases (36 bottles) were bad and wants ice wine in replacement. A little later she says she doesn't have any bottles as they were all thrown away.

The December credit card statement arrives at month

end and there is a dispute debit for this customer's purchase. I call the credit card company and explain the situation and our published return policy. They suggest I write a letter with all information. As I save everything, I include all communications with the customer. The dispute is reversed.

February: A new visitor arrives in the Tasting Room on Sunday and purchases a selection of wines. He says he works for the Wine Spectator magazine and would be interested in the Vinalies gold medal champagne. I offer one bottle for a review in the Wine Spectator. He will check with his editor. Later in the week, he leaves a message that they would be most interested in tasting the Vinalies gold medal champagne, but would not do a review, since their customers were not interested in Pennsylvania wines.

March: The ice wines are finished fermenting and can be bottled as soon as we can purchase bottles. Ice wine is the only product we make that the grapes can be grown, made into wine, and bottled for sale, in one year.

I receive an email from the Better Business Bureau in Bucks County about the Private Sale customer complaint. I communicate back with all of the details and the credit card company's decision. They cease and desist.

My left hand is becoming more of a problem. I have

to alternate the pruning days with days that I disgorge champagne. I prefer to finish the yearly champagne disgorging in the winter months when the cold weather makes the job easier. Now, I will go to a just-in-time approach throughout the year to give my hand a break. I'll use the coldest hours of the day in the early morning to finish a few cases at a time. This requires set up and clean up every day instead of once or twice a week.

For pruning, I use my right hand on the electric shears and usually clear the cut wood with my left hand. With the increasing pain in my left hand, I use my right hand to clear the cut wood, which slows me down, but I can still make progress.

In early April I get another email form the Better Business Bureau—this time in Lancaster, PA. It's the same customer complaint. Again, I respond with all of the communications including the credit card company decision. They persist that the customer has a new question about the Private Customer Sale wines. After some back and forth with this BBB, I say they have all of the information and communications regarding this matter, nothing is new, the credit card company has made their decision, and I consider this harassment on their part. I don't hear from them anymore.

I continue my efforts to disgorge champagne one day and

prune vines the next day. The barn swallows arrive around April 20th this year and keep me company in the vineyard. And the hummingbirds arrive on April 23rd. All is well.

May: The weather has been mild and has given me extra time to finish pruning. We have a new ice wine billboard on the Pennsylvania turnpike west of Valley Forge. It says "Ice Wine: More than a Kiss" with an artistic rendering of several of our bottles. We enjoy many new customers, especially from New Jersey. The New Jersey visitors are our favorites. They are open minded, friendly, and adventurous.

August: The barn swallows have been an amazing help in keeping all birds out of the vineyard, by patrolling the air space above the vineyard. Since 2012, I have left a window partially open in the tractor garage for them to nest. They can nest safely, feed on the insects from Jamie's mowing and around the area, and in return they patrol the vineyard. They chase all birds from the vineyard "airspace." We stand in amazement and watch them.

September: The white grape crop is small, but the juice at press is very flavorful. The Viognier and Gewurztraminer crops are too small for the half-ton press, so I pull out my old home winemaker fruit press. It still works well. I handle each batch with extra care, and I worry about the tanks not being full. I have a hand and wrist brace for my left hand now and can still lift the 40-pound picking lugs.

I blanket the air space in the containers with CO_2 to keep the liquid secure. (CO_2 is used to prevent microorganisms from growing on the surface of the wine *must*. These are particles which would impart unwanted flavors in the wine and could lead to spoilage).

Again, we decide to make the red varieties into ice wine. The customers can't get enough of the new red ice wines. Many customers bring friends, and some are interested in the dry wines. We continue to give a full tasting of the white wines, red wines, champagnes, and ice wines—that's ten to twelve ounces per taster—more than the legal limit of 8 ounces.

In October we offer three bottle samplers in special boxes of the white wines, the red wines, champagne, and ice wines for a 50% discount. We sell 250 boxed sets in two weeks.

I send another letter to the Township asking for the status of my application to sell the land development rights.

In December we order 500 holiday gift boxes for the samplers and sell out by mid-month. We're excited to see our customers all month, but the YTY sales are down 20%. At the end of December I send Armando home with half pay for January, with instruction to return to work in February. I'll have to start the pruning by myself in January.

2016

January: I can't use my left hand. I tie the thumb and forefinger together so I can use the other fingers. I try resting for two weeks but the pull of the vineyard is overwhelming. Jamie sharpens my electric pruning shears and helps me get the pack on. I have a brace on my left hand and can use the two middle fingers, I put a big mitten over my left hand, start pruning, and don't think about the next day. It's a good thing I am right-handed.

February: Jamie is in pain constantly and our doctor confirms he needs a hip replacement. He refers us to a surgeon in Reading. Our doctor's brother was a missionary in South America and the brother's plane went down in the Amazon. He was air lifted back to the States and was expected to be paralyzed. This surgeon in Reading operated and restored our doctor's brother's spine. Jamie says he doesn't want any more surgery—he had enough with the cancer. He plans to live with it.

Armando is back half time to help with the pruning.

We had to reduce his hours to conserve on the finances. I work on a new system of cutting back more to make the canopy smaller. This will eliminate one pass through the vineyard—so there will be only two passes to finish pruning.

Armando helps me in the early morning to move champagne cases to the work area and disgorge three cases on one of his weekly workdays. I'll do another three cases on the weekends to have finished inventory on hand.

One weekend in April a visitor from Lancaster County arrives and says her doctor told her to seek out wine without tannin. I explain that wine grapes as well as all pitted fruits have tannins, so all wine grape wines have tannins. Tannin in wine comes from the seeds and skins of wine grapes. She says that's not what her doctor told her. I reply that I've had a running battle with doctors for years about wine. They are not experts on wine.

I recall my many conversations with Dr. Donaldson about sugar and diabetes. Sugar is a no-no for diabetics, but I discover that I can consume both honey and wine grape juice with no problems. There must be a difference. Wine grapes have fructose and glucose sugars. The sugar formula $C_6H_{12}O_6$ is the same for sucrose (white sugar), honey, fructose and glucose, and other sweeteners; but the molecular structures are different.

We continue offering champagne by the glass and Saturday champagne-making tours at noon. For the first weekend in May we offer tours of the vineyard.

Jamie's hip is a problem for mowing with the Kubota mower. I use the John Deere brush mower to make the first cuts in the vineyards. The weather is mild and the under the vine growth is slow. The old vineyards have been groomed for years and have a mixture of clover, sweet grass, and wildflowers under the vine row, which won't be a problem. It never grows more than 20 inches tall and is well under the cordon wire. The new vineyards are still a little wild, so I'll have to do some weed control, unless Jamie is able to do the mowing with the Kubota.

June—I still prune a little each day. The buds are open and easily break off as I try to work around them. No choice—this is the way it is. Some crop will be lost, but the vines must be pruned.

Finally, Jamie can't stand the pain and we visit the surgeon in Reading. They explain in great detail the surgical procedure and the preparations. Jamie wants to wait as long as possible to decide. They offer a cortisone shot and say it may help. It doesn't.

We receive a letter from Warwick Township saying that we will be fined for growing Noxious Weeds on our property, as defined by the Pennsylvania Department of

Agriculture. The noxious weed list on the Department of Agriculture website reads like a tour of the Pennsylvania countryside. The roadsides are full of all of these plants. I determine that the only plant on the list that we always have is thistle—which is everywhere in Pennsylvania. Jamie can't sit on the mower, so I decide to weed spray by hand up and down the vine rows in the new vineyards. After ten days we can trim out the dead plants. Jamie is concerned that they will come to inspect. I tell him this is just harassment. These rows can't even be seen from the road anyway.

August: Pain wins out, and Jamie wants to get the hip replacement surgery ASAP. We work on the schedule with the surgeon. There is a cancellation in September. We are consumed with the preparations and tests before surgery.

Harvest is always on my mind but must take a back seat now. I decide that the entire vintage will go to ice wine. I won't have to press the frozen grapes until December. With my late pruning, the crop is small again.

September: Jamie's hip replacement surgery is a success. He did all the pre-op exercises and quickly makes progress with the physical therapist.

Armando picks grapes during his workday, and every other day I drive to the freezer after closing the winery. Denver Cold Storage is open until 7:00 p.m. and they leave a pallet on the loading dock for me. I can unload the 40 lb. lugs myself, and often the night manager helps me stack the pallet. He shrink-wraps it quickly and moves the pallet inside the cold room. Denver Cold Storage is a pleasure as they service many small farm operations in Lancaster County.

October: The grapes are all in the freezer. After six weeks the grapes will be as frozen as possible. Jamie is able to drive again and can handle customers in the Tasting Room using one crutch. We've had irregular hours for customers out of necessity and have not been able to do any events since the spring. I decide to see the hand surgeon, and schedule surgery for early January.

December: Armando and I have been to the freezer for the last 12 days and have the Ice wines pressed and fermenting. I send him home with half pay for January again. That's how we give him a little vacation.

Our attentions were elsewhere this year, and we end YTY revenue down 24%.

2017

January 5: We arrive by 7:00 a.m., to the Delaware County outpatient surgery center. Dr. Osterman will do the surgery on my left hand. The admittance clerk checks my form and asks about prescriptions. I say, "None." The pre-surgery nurse asks about prescriptions. I say, "None." My blood pressure is high. An intern checks my blood pressure again and it's a slight improvement. He asks about prescriptions, and I say "None." He replies, "Good for you!" Dr. Osterman comes to talk with me and looks at my left hand. As he walks away, I say, "Make it like original equipment!" He nods.

January is usually the slowest customer month of the year. This year we are surprised by many weekend customers. Jamie takes care of the customers and I stay in the back room doing whatever I can with my right hand. I carry bottles to the kitchen to clean them before labeling, and when Jamie has a break, he carries a case to wherever I'm working. We limp along with the labeling in a just-in-time-method.

Three weeks after surgery, the cast is partially removed from my left arm, and I begin physical therapy with a hand specialist. She reads the surgeon's report to me; they made new bones from the thumb knuckle to the wrist, new tendons to fasten to the wrist bone and repaired a nerve. She asks, "How did this happen?" I explain about the hand work required to complete the champagne making process. Specifically, the left hand moves the 3.5-pound champagne bottle four times to complete the dosage step. And the left hand puts the stopper on and off each bottle twice, with the thumb pulling the stopper off each bottle twice. The champagne is under pressure, so each pull is against three atmospheres pressure. Each disgorging session will work on between 70 bottles and 240 bottles. I've been working on champagne disgorging for 27 years. And there is also the bottling and labeling work which uses both hands to move the 2.5 to 3-pound full bottles. In one year, I would move about 6000 bottles through these two operations. And in addition, the vineyard and gardening work requires both hands.

After two weeks, she removes the remaining cast, and we start working on movement. I expect a quick recovery, but that is not to be.

I call Armando and explain about my hand and say he can work three days a week in February. He helps me in the winery label bottles and with the building cleaning. I can't do any pruning so I identify the old vine trunk sections that he can work on removing.

At the end of February Armando arrives with his family on Sunday and says he got another job with a landscape company in Norristown. They will pay him 28% more per hour, full time guarantee, two weeks paid vacation, and five paid holidays. If I match that he will stay. I say that I am sorry, but I can't do that. I wish him good luck. He says he can work on Saturdays starting next week.

March: I'm still confident that I will be able to start pruning soon. Physical therapy continues three times a week.

April: It's clear now that I won't be able to use my left hand for many months. We hobble along in the Tasting Room with Jamie taking all customers. We won't be able to offer any events or tours until I can use my left hand again. At night I think of how I can get to the most important vines so that we can have a crop this year.

May: The days wear on. I gaze at the vineyard and make plans on how to get the pruning done. The grass is growing and Jamie plans on getting to the Kubota to mow again soon. Armando is working overtime at his new job and hasn't been able to work on Saturdays.

I continue with physical therapy and start getting anxious about making progress. The hand specialist must report on the progress to the surgeon and asks about month-to-month change. I say I can use a nail clipper with the reconstructed thumb. "Good," she happily says. Later in the month she starts me using a two-pound weight to build wrist strength. That night I have intense pain in the wrist and thumb which lasts for five days. I cancel the next PT day. The next week she wants to continue with the weight work. I decide to stop physical therapy. I can't chance another episode of five more days when I can't do anything. I'll go it alone.

June: I can't stand it any longer. I have Jamie help me get the pruning shears pack on, and head to the vineyard determined to make some progress. I can use my right hand. The crop will be lost but they can be pruned back for the next year. When visitors ask about the vineyard, we say we had to cut back to let the vines recover from the Spotted Lanternfly damage.

The summer wears on as we feel that we are only able to perform at half speed in the winery and in the vineyard. We muster our energies every day and sigh with relief when the gate closes at 5PM. We work hard to present the image that everything is the same, that nothing has changed. Jamie tells the customers about my hand surgery and the work to have the vineyard recover from the pest problems. He takes on the difficult task of trimming the road bank flowers. We work from the roadside in to keep the appearances up. In July I can pick up a knife in my left hand. I start weeding the gardens in August. It takes until December to make them look groomed.

We offer our customers to picnic in the vineyard during the fall, as we can't offer any events.

The year ends down 28% YTY.

2018

January: After the annual Federal Excise Tax and BATF (Bureau of Alcohol, Tobacco, and Firearms) forms are filed, I decide to have a garage sale inside the winery. I gather up framed prints, an antique dish set, a Navaho pottery collection, vases, crystal serving bowls, porcelain platters, candlesticks, caned chairs, antique champagne flutes, braided rugs, and Mexican pottery, and set them up for sale in the Tasting Room. We have open house weekends starting in late January through March for customers and take them on a walking tour of the interior winery spaces. Everyone seems to enjoy the activity. The vineyard is dormant, so visitors don't expect to see the grapes growing. We revive the multi-bottle discounts: 20% on three bottles, 30% on six bottles, and 45% on twelve bottles, mix and match. And champagne by the glass is available on the weekends.

March: Last fall I completed the pruning cuts in the old vineyard. The cut canes have been hanging in the wires waiting to be removed. My left hand is reasonably functioning now, so I start to remove the cut canes and collect them for burning. I push my garden cart down the row and when the cart is full, I push the cart to the burn pile. One row a day is slow, but progress is made. A fellow vineyard grower has a worker that can work on Saturdays. He can clear the cut canes from four or five rows per day. During the week I use my pruning shears to refine the cuts I made last year. I purposely cut back hard to the top of the vine trunks, looking for green tissue. The green tissue signifies that the vine is alive at that place and should sprout new growth from that place. This will keep them viable for the future.

May: The indoor garage sale worked well so I plan on an outside version starting on Father's Day weekend. We have extra furniture stored in the barn and the house and move everything outside. I set up tables for tools, garden implements, and garden decorations. It's a beautiful day and many "fathers" visit. We offer a complimentary glass of champagne for all "fathers." We can't offer the big events anymore, but these small activities are well received.

June: I notice a new real estate office on Route 23 near Route 100, about five miles away, with a sign "Wanted - Land for Sale." I talk with Jamie about the idea to sell the second parcel. That would clear the mortgages and allow us to stay at the original 13-acre parcel, with the original vineyard and all the buildings.

In 2004 the Township committed that I would get their support to "sell the farm's development rights," if I purchased the second parcel of 19 acres and farmed it—which I did and have been doing for 14 years. This second parcel combined with the original parcel of thirteen acres would in total qualify for the grant. But they never "approved" my application. So I had to pay for a second mortgage to cover the purchase price of the second parcel for these 14 years, instead of the help I needed to farm the land.

In Pennsylvania the Agricultural Easement Program was started in 1988 to slow the loss of farmland to non-agricultural uses. The program enables state, county, and local governments to purchase conservation easements, sometimes called development rights, from owners of farmland. The state gives funds to the counties to distribute. When a farm owner "sells" the farm's development rights, the land is "restricted" to farming use only. In return the farm owner receives cash per acre. The cash amount is the county's estimate of the reduced sale value of the land as farmland versus land to be developed. The farm is then deed restricted.

This realtor visits to discuss the possibilities of selling the second parcel of 19 acres.

July: The Spotted Lanternfly invasion is in full swing in Eastern Pennsylvania. We notice them around the buildings, but not in the vineyard.

August: The "Land for Sale" realtor stops by and says that the 19-acre parcel is beautiful but people are looking for one-half or one-acre parcels to build on. He has one buyer looking for a larger parcel, but they wanted to know if the vineyard could be removed. We were surprised as we never thought anyone would want that.

September: Everyone asks us about the Spotted Lanternfly and what to do. We've seen our praying mantis catch them and have learned that a European wasp also will feed on them. Many vineyards report problems. I ponder why we don't have that problem.

October: We offer a huge mix and match case (12 bottle) discount to boost annual sales and move inventory. I can finish champagne disgorging again by myself, and work on four cases/day.

A woodworker friend agrees to make cutting boards from the milled black walnut wood that's been "drying" in the barn. The cutting boards for sale are a big hit. I invite several wood workers to set up in the upstairs of the barn with items for sale and offer the black walnut for sale at a bargain board foot price to our customers. We're flooded with customers, and they come back all fall for hardwood boards.

November: I talk with some local farmers about removing the vineyard. One farmer knows about a couple of home winemakers looking to start a vineyard. They are interested in salvaging the 20-gauge steel wire and vineyard posts. They start in one of the new vineyard blocks in the 19-acre parcel.

December: We invite the local artists group to display their work for sale in the Tasting Room in December and host a "Meet the Artist" champagne reception on the first Saturday afternoon.

The "Land for Sale" realtor stops by with a developer friend to look at the 19-acre parcel. Neither of them has done any work in our Township, and they plan to stop by the Township office next. They are not optimistic.

We finish with flat YTY sales.

2019

January: The weather is mild and the vineyard salvage work continues. Through word-of-mouth there is a lot of interest in the vineyard posts. These specialized southern pine treated posts retail for $10-$15 each. We only ask for help with the work of taking the posts out of the vineyard.

The thought of selling this 19-acre parcel is with mixed emotions, but the prospect of keeping the original vineyard and the winery outweighs any regrets.

Our friend Andy helps with the problem of wire removal. He takes my spinning jenny wire tool and fits it in a stand. A spinning jenny is used to feed the 20-gauge wire off the spool. This stand allows the spinning jenny to spin easily in reverse as I pull the wire out of the vineyard. I can pull two wires at a time—that's 250 feet of wire onto the spool. Then a zip tie secures each spool. The spools and the rebar vine stakes are stored in the old green van for later transport to the Meyer Pollock scrap metal company in Pottstown. We would take about 1500- 2000 pounds per trip to Meyer Pollock and get 2-4 cents per pound.

February: We decide to contact realtors about listing the 19-acre parcel for sale. We pick three different realtors and arrange on-site visits. They all ask about the winery property and recommend listing both parcels, with a premium on the winery property. They say that will generate more interest in the 19-acre parcel. We decide to take a time-out and consider our options.

It remains a desirable option to sell the 19 acres so we can continue the winery. We have not had a harvest for two years, but the vines are ready to produce any time. The lack of harvest has reduced expenses and made it easier to cover the other operating costs. We know from friends in the business that it takes a long time to sell a vineyard and winery—several years on average. It may make sense to follow the realtors' suggestion and list both parcels for sale, in the hopes of a buyer for the 19-acres. We get busy on the listing paperwork and pictures.

March—There are a lot of empty barrels, which are hard to keep sterile and hydrated when not in use. I decide to offer them for sale to our customers, and it's a big hit. And most of the barrels are French oak—more enticing to customers.

April: Prospective buyers visit, but most are interested

in the thirteen acre winery and vineyard—not the nineteen acre parcel. This takes more of our time than expected. The barrel sales continue strong and bring winery customers back. With the reduced winery production in mind, I look for other equipment that can be sold.

June: The township administrator leaves a phone message that my application to sell the land development rights has been approved. It was fifteen years ago that they said they would support my application to sell the land development rights for my two parcels, if I bought Lester's 19 acres. The purchase of Lester's land qualified my lands for this program. The program is designed to keep agricultural land from being developed.

The money I would have received would have paid for most of the cost of purchasing the new land. Instead, I had to get a second mortgage to pay for the land. For fifteen years, I have paid that second mortgage, instead of paying additional workers to help me farm the land. This would have been a difference maker. Now I have to sell that land to keep the original 13-acre vineyard and winery property.

August: Half of the bird netting and the picking lugs are sold by advertising in the Penn State Department of Agriculture newsletter.

September: The weather is cool enough to start

disgorging champagne again. I set up for a tour on Saturdays for customers of the champagne making process including the disgorging and corking steps. The YouTube video has the complete process, but most people have never witnessed this "live" It's a big hit. Everyone especially likes the popping of the champagne bottle cap and the fast dispersion of the sediment.

November: We've fielded many visits from prospects over the last six months. Most express interest in buying the winery and vineyard. After spending time touring the property, the prospects see that their dreams don't match the reality.

I recall the surprise of a customer after going on a vineyard tour, "This is really farming!" The "dream" of owning a winery rarely is accompanied with knowledge of the industry. The vineyard is agriculture, the winery is manufacturing, and the business is retail and distribution.

December: Wine sales continue slowly, and prospective buyers keep visiting. Twenty-eight champagne storage boxes are available for sale. That's over 10,000 champagne bottles that have been sold over the past five years.

The year ends down 25% YTY.

2020

January: There are several prospects that we are discussing options with. Everyone wants help with the vineyard and winery. I offer one year of assistance—a full calendar year of help that will include all the aspects of work. We are well along in our thinking that we will have to sell the vineyard and winery property—hopefully to someone who can continue it.

March: The Pandemic takes over the country.

April: The Commonwealth of Pennsylvania closes all wineries, restaurants, gyms, and other businesses not deemed essential.

June: Pennsylvania opens up their state stores for alcohol sales, but the wineries, restaurants, and other non-essential businesses remain closed.

July: Pennsylvania notifies the wineries that with the purchase of an additional license and a health department certificate, wineries can serve tastings of wine in approved outside seating only, and food must be served. This is nearly impossible and requires more expenditures.

September: The Pennsylvania Department of Revenue sends a letter asking for copies of all of our receipts for expenditures on the 2019 Federal tax return. We contact our accountant and discover that the state has sent this letter to all small businesses. The current year revenues are significantly down due to the Pandemic restrictions— and it appears they are looking for ways to raise revenue. They have no jurisdiction over the Federal tax returns of a business.

October: We advertise equipment for sale in the wine business publications, and immediately get contacted and have winery and vineyard industry visitors from New York, Connecticut, New Hampshire, New Jersey, Pennsylvania, Maryland, Delaware, Virginia, West Virginia, North Carolina, Ohio, Michigan, and Indiana. We hear from these visitors: "We heard about you and always wanted to come here!" All these visitors ask to see the champagne award. They ask how I made the famous champagne. It feels like they are truly interested and appreciate anything I can tell them. I answer all their questions and listen to what they are doing with their wine and vineyards. I am honored by their visits, and the recognition by my peers that through unrelenting hard work we achieved our goals.

November: We offer a private wine liquidation sale to our customers. Technically this is not allowed by the state, but we need revenue.

December: The future for the winery is uncertain, but we continue paying the license fees for the next year.

2021

The Pandemic rages on.

Businesses are closed.

So are we.

I'll be 71 years old this year and Jamie is 74. I'm probably 10 years too old to be thinking about continuing.

The will is still strong, but the body is broken.

Keeping up appearances is no longer possible.

We are blessed with someone who will buy the 32 acres including buildings.

We retire to South Florida to start the rest of our lives–literally.

The End

Afterword

A friend asked me, "Would you do it again?"
"The winery?"
"Yes," she replied.

I thought about how the answer is that I would have to evaluate the many, many decisions in my life that came before the winery and vineyard. As a young person, I didn't know that a vineyard and winery existed. After college, I learned about champagne and had the chance to make wine with friends from college. My grandmother gave me a glimpse of the beauty of nature with her house plants, and I started my own plants. I liked honey and started to keep bees. As soon as I had a back yard, I started a tomato garden. I was interested to learn how things worked and tore the engine apart in my used VW. I learned medical research laboratory techniques at the Veterinary Hospital, and business and management skills at IBM. I wanted to learn and do everything, and believed I could.

Never would I have imagined the contentment of being that comes from working and striving with nature.

Vineyard and winery work is hard and unyielding. The results are not evident for many years. This idea of longevity is not usually in a young person's dream. If my late husband's life- threatening diagnosis had occurred in the early days of the winery and vineyard, this endeavor would most likely have ended then. In my first book *Time and Tides*, these challenges are evident. We had been operating the vineyard and winery for 12 years, so we continued.

But it was a daunting task to go it alone. My life's experiences said, "Just continue." Just do my best every day was how I had lived the 53 years before Fred passed away. And so, I continued. After 30 years of small "family farming," life's forces eventually win out. And the Pandemic was a dam that couldn't be forged.

I'm blessed with my wine, the Vinalies awards, and wonderful memories of the special people who helped me along the way.

Would I do it again?

Of course I would!

In a heartbeat.

www.ingramcontent.com/pod-product-compliance
Lightning Source LLC
Chambersburg PA
CBHW020350170426
43200CB00005B/115